D1328779

Genetic Engineering

Genetic Engineering
Catastrophe or Utopia?

Peter R. Wheale
Director, Bio-Information (International) Limited,
Chairperson, Biotechnology Business Research Group and
Senior Lecturer in Management and Business Studies,
Oxford Polytechnic School of Business

Ruth M. McNally
Director, Bio-Information (International) Limited and
Visiting Honorary Fellow, Oxford Polytechnic

HARVESTER · WHEATSHEAF · ENGLAND

ST. MARTIN'S PRESS · NEW YORK

First published 1988 by
Harvester • Wheatsheaf
66 Wood Lane End, Hemel Hempstead
Hertfordshire HP2 4RG
A division of Simon & Schuster International Group

and in the USA by
ST. MARTIN'S PRESS, INC.
175 Fifth Avenue, New York, NY 10010

Printed in Great Britain by Billing and Sons Limited, Worcester

British Library Cataloguing in Publication Data

Wheale, Peter
Genetic engineering : catastrophe or Utopia?
1. Genetic engineering
I. Title   II. McNally, Ruth
575.1   QH442
ISBN 0–7450–0010–X

Library of Congress Cataloging-in-Publication Data

Wheale, Peter.
Genetic engineering.
Bibliography: p.
Includes index.
1. Genetic engineering—Social aspects.
2. Genetic engineering—Economic aspects.
I. McNally, Ruth.   II. Title.
QH442.W49 1988   575.1   86–27915
ISBN 0–312–00479–6

1 2 3 4 5 92 91 90 89 88

In loving memory of
Patrick McNally and John Harry Wheale

Contents

List of Figures

List of Tables

Preface

Science and technology are Janus-faced, simultaneously generating catastrophic as well as utopian scenarios.

Daedalus, the fabled Athenian artificer, was ascribed with the invention of such tools as the axe and the saw. It was he who first fixed arms and legs to the *xoana*, the shapeless primitive statues of the gods. Daedalus also constructed the Labyrinth, an amazing palace from which no one could find an exit, and it was in the Labyrinth that the Minotaur, a flesh-eating monster, half-human, half-bull, was enclosed by King Minos of Crete. Having offended King Minos, Daedalus and his son, Icarus, were imprisoned in the Labyrinth, but Daedalus, with his ingenuity and cunning, devised a pair of wings which enabled him and Icarus to escape. However, in flying with his father from Crete, Icarus soared too near the sun and the wax by which his wings were attached melted and he plummeted into the sea.

The genetic engineer, like a contemporary Daedalus, claims to be providing society with a vast range of innovations, such as more effective and cheaper pharmaceutical products, more abundant food crops, new approaches to the generation of energy, the recovery of resources and pollution control, and the diagnosis and correction of genetic disorders.

On the other hand, as a result of the application of genetic engineering, worldwide pandemics caused by newly created pathogens, the triggering of catastrophic ecological imbalances by the release of novel organisms into the environment, the creation of new agents of biological warfare and the increased power to manipulate and control people, may each become realities in the near future.

This book is intended for students of science, social studies and business, and the lay reader interested in understanding and evaluating the effects that the revolution in genetic engineering is having on our society. It is our hope that it will contribute to an improved dialogue between the individuals comprising these different groups.

No specialist knowledge in the sciences or the humanities is required to understand the ideas, arguments and discussions presented.

An extensive bibliography, a comprehensive glossary of technical terms, name and subject indices, and a compendium of educational resources for genetic engineering and biotechnology are provided at the end of the book.

Peter Wheale and Ruth McNally

April 1988

Oxford

Nature is so fecund that any careless attempt to alleviate poverty will encourage unsupportable increases in population and would thus only exacerbate the suffering it is designed to relieve. As far as I am concerned, nature is unimprovable. Social reformers should therefore allow events to take their inevitable course and let war, disease and starvation reap the surplus.

Thomas Malthus, 1798

Whatsoever, in nature, gives indication of beneficent design, proves this beneficence to be armed only with limited power; and the duty of man is to co-operate with the beneficent powers, not by imitating but by perpetually striving to amend the course of nature—and bringing that part of it over which we can exercise control, more nearly into conformity, with a high standard of justice and goodness.

John Stuart Mill, 1885

Our business here is to be Utopian, to make vivid and credible if we can, first this facet and then that of an imaginary whole and happy world. Our deliberate intention is to be not, indeed, impossible, but most distinctly impracticable by every scale that reaches only between today and tomorrow.

H.G. Wells, 1905

Part 1
Revolution in Genetics

1 The Double Helix

When future generations look back and attempt to characterize the twentieth century they will surely say it was above all else the Age of the Atom. The intellectual shock waves of the development of quantum physics have had a profound effect on the whole of twentieth-century life. Nowhere is this more evident than in biology. Physicists brought to biology their powerful atemporal reductionist methodology — the belief that living phenomena can be understood in terms of the laws of physics and chemistry — the legacy of which is molecular biology. Molecular biology is a quantum approach to the study of living systems which endeavours to understand biological phenomena in terms of molecular structures, and all fields of biology have been profoundly influenced by it.

Molecular genetics was made possible by the confluence of theoretical and experimental advances in classical genetics, biochemistry and molecular biology. Whereas classical genetics was concerned with the problems of evolution, speciation and phylogeny, molecular and biochemical genetics constitute an atemporal approach to the control systems of cells.

Genetics is a term that was coined in 1906 by William Bateson (1861–1926), the 'apostle of Mendelism in England' (see Magner 1972). It is derived from the latin word *genesis* pertaining to the origin or generation of a thing or the mode of it. The science of genetics endeavours to resolve two apparently antithetical observations—that organisms both resemble their parents and differ from them. Genetics, in a single theory, down to the molecular level, attempts to explain in one synthesis both the constancy of inheritance and its variation (Lewontin 1984).

During the first thirty years of the twentieth century quantum mechanics developed an understanding of atoms and subatomic particles and scientists began to appreciate how complex molecules are formed and held together. The broad theoretical perspectives of quantum physics and chemistry were developed primarily in Central Europe. Social and economic conditions prevailing there, particularly in Germany in the 1930s, motivated the exodus of a large number of people, including academics, a high proportion of whom eventually settled in the USA. By virtue of their theoretical perspective, certain emigré scientists were to make significant contributions to newly developing inter-disciplinary scientific specialities, many of which were relatively underdeveloped in US universities, such as theoretical physics, applied mathematics, quantum chemistry, biophysics and biochemistry, and what later came to be called molecular biology (Coser 1984). Displaced from their country of origin and often peripheralized in their own discipline, a number of physical scientists left the occupation for which they had been trained and focused their attention on the elucidation of the three-dimensional structure of biological molecules (Fleming 1968).

The rise of molecular genetics was inspired by the speculations of several eminent quantum physicists on the nature of the hereditary molecule. In 1933, following his formulation of the quantum theory of atomic structure, Niels Bohr (1885–1962), the Danish nuclear physicist, hypothesized that it may not be possible to account for certain biological phenomena wholly in terms of conventional physical and chemical explanations (Frank 1949). Bohr's hypothesis was refined by his pupil Max Delbruck (1906–81) in a paper entitled, 'On the Nature of Gene Mutation and Gene Structure', published in 1935, in which he stated that a domain of biological enquiry in which explanations in terms of conventional physics might turn out to be 'insufficient' in Bohr's sense was genetics (see Stent 1971). Erwin Schroedinger (1887–1961), the

Austrian theoretician of quantum mechanics, popularized these views in his book *What Is Life?* (Schroedinger 1945).

In 1953 a physical basis for the gene in the structure of the deoxyribonucleic acid (DNA) molecule was proposed by the American James Watson, a member of the so-called 'phage group' and the physicist Francis Crick working at the Laboratory of Molecular Biology in the Cavendish physics laboratory in Cambridge. Phage (Gr. *phagein*: eat.), or bacteriophage, are viruses which infect bacteria. The phage group was a closely knit group of innovative scientists at different universities and research establishments in the USA, many of whom, like Delbruck, had been trained in the physical sciences. What distinguished the members of the phage group most sharply from their predecessors in phage research was their single-minded interest in resolving the physical basis of heredity and in solving the problem of how a chemical could encode genetic instructions (Stent 1971).

The chemical DNA had been discovered as early as 1869, at the time of the emergence of the new science of biochemistry, by the Swiss chemist Johann Friedrich Miescher (1844–95). Miescher identified 'nuclein', as he called it, whilst he was studying the physiological chemistry of the cellular nucleus. Despite evidence that it was able to convey and to alter heredity, for most of the first half of this century DNA was not taken seriously as the chemical basis of the gene. This rested to some extent on the inaccurate 'tetranucleotide hypothesis' of the 1920s, the implication of which was that the structure of the DNA molecule was too simple to encode the requisite vast amount of hereditary information, and also on the importance which had been attached to the structural complexity of proteins.

Proteins are a diverse class of molecules including enzymes, hormones, structural proteins such as skin, muscle, hair and nails, and blood proteins such as haemoglobin. Proteins form the basis of the structure of living organisms and the metabolic processes characteristic of life are dependent upon their function. Each different protein is composed of a precise sequence of amino acids. Contrary to the stance of the majority of chemists, physicists and geneticists of the time, the major tenet of the phage group was to act on the evidence that the chemical nature of the gene was DNA rather than protein.

Within the DNA molecule there are chemical groups called nucleotide bases, of which there are four types: adenine, thymine, cytosine and guanine. In 1950 Erwin Chargaff proposed the equivalence rule for the structure of DNA, which was based on observations that the ratio of the amounts of the chemical bases adenine to thymine, and cytosine to guanine, are always very close to unity for DNA molecules from a variety of organisms. This observation, incorporated into the model-building research of Francis Crick and James Watson, countered the notion of the tetranucleotide hypothesis that the structure of DNA was not complex enough to be informational (Watson 1968).

The technique of X-ray crystallography was crucial to research on the structure of DNA. X-ray crystallography is a method of obtaining photographs of diffraction patterns of the atomic structure of crystals which are used to visualize their three-dimensional structure. It was developed by a group of physical scientists whose aim was the understanding of the function of molecules through their structure. After the Second World War a team of researchers led by Maurice Wilkins, a physicist who had worked on the Manhattan project which developed the atomic bomb, was established at the Biophysics Unit of King's College in London with the remit of investigating the structure of DNA using X-ray crystallography. In order to assist Maurice Wilkins' team with the complex process of interpreting the X-ray diffraction photographs in terms of three-dimensional structures, Rosalind Franklin, an experienced X-ray crystallographer, was appointed to King's College in 1951.

Early in 1953 Wilkins supplied Watson with a print of one of Franklin's X-ray diffraction photographs of DNA which was far superior to the X-ray photographs of DNA with which Watson and Crick had been working previously. Watson and Crick interpreted the diffraction pattern on Franklin's print as indicating that the structure of DNA was a regular helix, the measurements of which indicated that it was a double helix; that is, it was a molecule composed of two strands twisted around each other. In April 1953, following a short period of intensive work on building scale models of DNA with almost every conceivable arrangement for the pairs of bases except the correct one, Watson and Crick finally proposed the famous double helical structure for the DNA molecule for which they shared the Nobel prize for Physiology and Medicine with Maurice Wilkins in 1962 (see Watson 1968; Olby 1974; Sayre 1978; Judson 1979; Gribbin 1985).

The key structural features of the double helix DNA molecule are the variation embodied in permutations in the sequence of the bases—adenine, thymine, cytosine and guanine—and its duplex structure (Watson and Crick 1953). These features endow the nucleic acid, DNA, with the requisite properties of a genetic molecule, namely the capability to record a very large number of genetic instructions in a manner which can be passed from generation to generation.

The DNA double helix is a double-stranded molecule which may be visualized as a very long spiral step ladder with thousands of millions of rungs (Fig. 1.0.1a). The sides of the ladder are formed from alternating sugar molecules (deoxyribose) and phosphate groups (Fig. 1.0.1b). To each sugar molecule is attached one of the four bases. A base plus a sugar molecule plus a phosphate group is called a nucleotide, and DNA is a polynucleotide, that is a molecule comprised of many nucleotides.

The nucleotide bases, adenine, thymine, cytosine and guanine, are commonly referred to by their initial letters as A, T, C and G, respectively. The rungs of the ladder are formed from pairs of bases, the pairing of which obeys the complementary base-pairing rule, which is that A always pairs with T, and C always pairs with G (Fig. 1.0.1c). The Watson–Crick structure can accommodate any sequence of base pairs along the double helix, and it is the sequence of bases which encodes genetic information (Fig. 1.0.1d).

The amount of DNA within each human cell is enormous and corresponds to a total length of about 2 metres. There are approximately 3×10^{12} cells in the human body. Since there are about 2 metres of DNA in each cell, this means

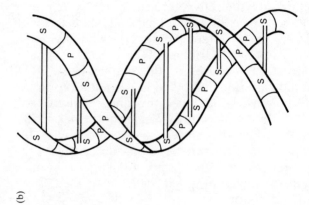

(a)

The structure of the double-stranded DNA molecule may be compared to a very long spiral step ladder with thousands of millions of rungs.

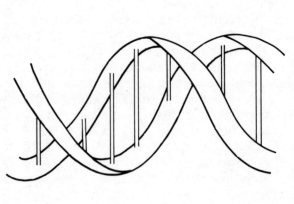

(b)

S = Sugar (deoxyribose) molecule

P = Phosphate group

The sides of the ladder are formed from alternating sugar molecules (deoxyribose) and phosphate groups.

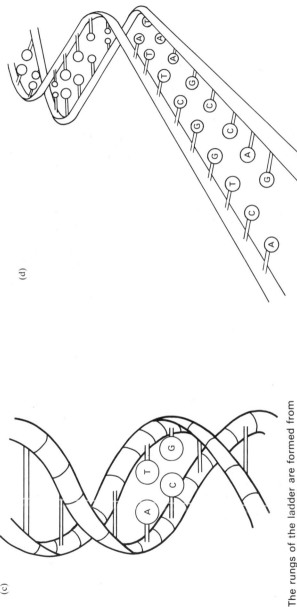

(d)

The sequence of bases in the DNA molecule encodes hereditary information.

(c)

The rungs of the ladder are formed from pairs of bases, coupled in accordance with the base-pairing rule:

base A (adenine) always pairs with base T (thymine); base C (cytosine) always pairs with base G (guanine).

Figure 1.0.1 The double helix

that a human has about 6×10^{12} metres of DNA. Joined end
to end this DNA would reach to the moon and back nearly
8,000 times (Weatherall 1985).   To fit 2 metres of DNA
into the tiny space within each cell, the DNA is packaged
into the distinct structures known as chromosomes. A
chromosome is a double-stranded DNA molecule supercoiled around a core of protein molecules (Fig. 1.0.2). In
human body cells there are forty-six chromosomes, composed of twenty-three pairs. Chromosomes differ in size,
but on average each of the forty-six chromosomes contains
approximately 150 million DNA base pairs.

Cells multiply by a process called cell division in which
a cell splits into two and makes two daughter cells. Prior
to cell division, the cell makes a duplicate of the genetic
material in its chromosomes so that each daughter cell
receives a full and accurate set of genetic information; that
is, it replicates its DNA. The model for the structure of
DNA proposed by Watson and Crick was compatible

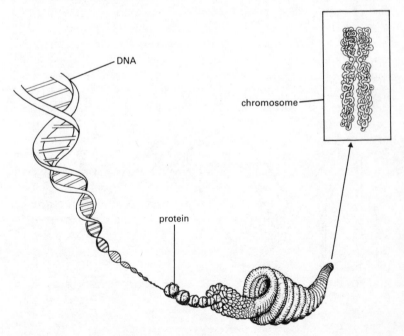

Figure 1.0.2 Chromosome structure

with the faithful replication of genetic information, thus preserving it through successive cell divisions. The key feature of the double helix structure in this respect is the complementary base-pairing, whereby the bases always pair A with T, and C with G, as a result of which the two halves of the long duplex DNA molecule are complements of each other. The consequence of complementary base-pairing is that one of the two strands of a DNA molecule can provide a template for the assembly of the base sequence of the other, complementary strand of it.

The union of the two complementary halves of the double helix along the centre line is fragile and during replication the DNA molecule splits along this centre line, producing two non-identical complementary halves. Each half serves as a pattern, or template, upon which the base sequence of the original double strand is reconstituted, and thereby two identical double-stranded daughter DNA molecules are produced.

Crick was instrumental in both formulating and elucidating the principles of a discipline which perceived of genetics as a molecular subject. In 1957 he presented a paper to the Society for Experimental Biology in which he elaborated upon the state of knowledge in molecular genetics and 'puzzle-solving' lines of research that could be fruitfully pursued (Crick 1958). This paper was to have a direct influence on future research orientation in molecular genetics (see Carlson 1966). In Crick's view, the major problem of molecular genetics which required investigation was how the genetic information encoded in DNA was translated into protein molecules. In his paper Crick stressed the importance of what he termed the central dogma of molecular genetics which states that there is a uni-directional flow of genetic information from nucleic acids to protein. The central dogma holds that once information has passed into protein, it cannot get out again. The central dogma became a guiding principle in subsequent research to isolate and identify the components of the 'genetic machinery' of the cell which 'translates' genetic information encoded in DNA into protein molecules.

Proteins are composed of long folded chains called polypeptides which comprise of between fifty and more than 2,000 amino acids. Approximately twenty different amino acids are used in the synthesis of proteins by living organisms, and the proportion of a given amino acid varies from one protein to another. The genetic code deciphers the relationship between the sequence of bases in a DNA molecule and the sequence of amino acids in the protein for which it codes.

That there was a relationship between genes and enzymes, a class of proteins, had been suggested fifty years before, at the turn of the century, by the British physician Archibald Garrod (1857–1936), to explain 'inborn errors of metabolism', a class of inherited disorders (Garrod 1963). The dictum coined in the 1930s, 'one gene — one enzyme', refined the concept, expressing the hypothesis that each gene produces a single specific initial effect by the production of a single specific enzyme. The Watson–Crick model of a double helix structure for the DNA molecule with its genetic message encoded in permutations of four bases was compatible with the prediction that a gene directs the assembly of amino acids into chains to form proteins.

By June 1966 the genetic code was fully deciphered. The basis of the code is virtually universal throughout living organisms. Each amino acid is coded for by a triplet of bases which is called a codon. There are sixty-four possible codons to code for twenty amino acids, therefore more than one codon can correspond to a single amino acid.

DNA resides in a membrane-bound compartment of the cell called the nucleus. Amino acids are assembled into polypeptide chains outside of the nucleus, in the cytoplasm. The existence of an intermediary molecule, which would convey the genetic message of DNA into the cytoplasm to direct the assembly of amino acids into proteins, was predicted before its isolation in 1960. This intermediary molecule was given the name of messenger ribonucleic acid (mRNA). The major differences between the two nucleic acids, DNA and RNA, are that the sugar molecule in RNA is ribose, rather than deoxyribose, and in RNA the base thymine (T) is replaced by the base uracil (U). Transcription, the synthesis of the intermediary mRNA, occurs by complementary base-pairing on a single strand of DNA.

The central dogma was a guiding principle of research

into the elucidation of the genetic code. It states that there is a unidirectional ·flow of information from nucleic acids to proteins. The central dogma of molecular genetics uses the metaphor of language: for gene expression to occur, information encoded in the language of the nucleic acid DNA is transcribed into the language of mRNA which is then translated into the language of proteins. The central dogma is more than just an extreme restatement of the germ-plasm theory, postulated in 1883 by the German naturalist August Weismann (1834–1914) at the time of the so-called Lamarckian backlash in evolutionary theory.

Jean-Baptiste de Lamarck (1744–1829) proposed the first comprehensive theory of natural evolution in 1809. Central to his theory of evolution was the idea that the natural habits of creatures and the environmental influences upon them would inevitably lead to hereditary modifications of their anatomical structures. Lamarck called this process the inheritance of acquired characteristics. Towards the latter part of the nineteenth century there was a revival of the theory of the inheritance of acquired characteristics.

In the germ-plasm theory, Weismann used the concept of the isolation, purity and inviolability of the germ-plasm to argue against theories of the inheritance of acquired characteristics. Germ-plasm is a term denoting that part of an organism which passed on hereditary characteristics to the next generation. According to Weismann, the germ-plasm was transmitted from generation to generation through the germ-line, the cells from which sex cells are derived. Weismann pointed out that in animals the germ-line is set aside early in embryological development and he hypothesized that since the germ-plasm is isolated from the body cells, a change in the latter as a result of experiences and environmental influences during the lifetime of the animal, cannot affect the former. Thus, according to Weismann's germ-plasm theory, the transmission of characteristics acquired during the lifetime of an animal is not possible. The central dogma is not just a refutation of the inheritance of acquired characteristics, but a statement of the primacy of nucleic acids above other cellular components, in all cells not just those of the germ-line.

After the central dogma of molecular genetics was

elaborated, the consensus amongst molecular biologists was that genetic information flows unidirectionally from DNA to RNA to protein. However, in 1970 it was independently discovered by two groups of researchers in the USA that the replication of various RNA tumour viruses, called retroviruses, involves the passage of genetic information from RNA to DNA, in contradiction to the dictates of the central dogma. At first this discovery was deemed a serious challenge to the central dogma of molecular genetics. To accommodate this apparent anomaly the central dogma was reformulated to state that once information had passed into protein it cannot get out again. It was postulated that the transfer of information from nucleic acid to nucleic acid, or from nucleic acid to protein, may be possible, but the transfer or information from protein to protein, or from protein to nucleic acid is impossible.

In 1866 Gregor Mendel (1822–84), founding father of modern genetics, published evidence that certain hereditary traits are controlled by particles which are transmitted from parent to offspring as unit particles. The concept of the discrete hereditary particle, now known as the gene, has since taken on a significance in both scientific and popular thought about living organisms that far exceeds the role for which it was originally introduced. The gene is credited with controlling heredity, and by extension metabolism, appearance, behaviour and nearly every other characteristic of living organisms (Lindegren, 1966).

The resolution of the investigations into the genetic code and of the mechanisms which translate genetic information into proteins by the adoption of a linear sequential model has reinforced the reductionist approach to genetics. However, even at the molecular level, such a model is known to be simplistic. DNA contains many different kinds of message encoded in many different languages, and the triplet genetic code deciphered just one of them, the relationship between the linear sequence of bases in DNA and the linear sequence of amino acids in the polypeptide chains of proteins. Other kinds of information encoded within DNA relate to the control of gene expression, DNA replication, transcription and chromosome segregation. For example, an AT-rich region known as the 'TATA box' is believed to help to direct enzymes to the correct initiation site for transcription. Although the sequence of bases in the genetic code are arrayed in a linear fashion, the DNA molecule is a coiled molecule. Even at the time the model for the structure of DNA was postulated it was known that the double helix was a simplistic model of the structure of the genetic contents of cells. DNA is not found as a free molecule in eucaryotic cells but is organized with protein molecules into chromosomes, which represent a higher order of organization than the linear sequence of bases in the DNA molecule (see Fig. 1.0.2.). The three-dimensional coiled structure of DNA and the supercoiled structure of chromosomes create complications for the simple models of DNA replication and the transcription of RNA molecules.

Although the double helix model for the structure of DNA is able to account for how the molecule encodes genetic information, it does not reveal how the expression of that information is controlled. Each cell in the human body contains a full set of chromosomes. In any given cell, however, just a small amount of the total genetic information is relevant to its specialized role within the body and that genetic information is selectively expressed. Thus, within a cell the relative amounts of gene products varies enormously from gene to gene indicating that there exist processes which control the rate of gene expression from DNA through mRNA to protein molecules.

The first insights into the control of gene expression came not from molecular genetics studies on microbes, but from classical breeding experiments on maize plants conducted in the 1940s by the American geneticist Barbara McClintock (born 1902) at Cold Spring Harbor in the USA. In the early 1950s McClintock began to call attention to the behaviour of certain DNA sequences in maize that she called 'controlling elements'. These controlling elements, which were first noticed because they inhibited the expression of maize genes with which they came into close contact, did not have fixed chromosomal locations. Instead they seemed to move about the maize genome (Fedoroff 1984; Keller 1983).

Fundamental changes in scientific thinking, according to the American historian of science Thomas Kuhn, occur when sufficient 'anomalies' between fact and theory have accumulated to cause a 'paradigm crisis', which eventually results in the emergence of a new paradigm (Kuhn 1970). A paradigm is a research group's constellation of beliefs, values and scientific techniques. However, the dissemination of relevant information important in determining what is to be considered legitimate scientific knowledge is mediated by professional groups.

McClintock's announcement of her discovery of regulatory mechanisms operating at the genetic level presented a challenge to the central dogma of molecular genetics because it posed the question: what controlled the regulators? (see Keller 1983). Her results were not favourably received by the mainstream scientific community and her work on controlling elements in maize was for many years peripheralized. The concept of 'jumping genes' — elements within the genome that can move about — ran counter to the consensus of beliefs amongst her contemporaries who had adopted the concept of a stable genome developed by certain physicists and chemists. McClintock's experimental material and her classical pre-molecular methodology differed from those of mainstream geneticists. This difference distanced her work from their work and also served to undermine the credibility of her results as perceived and her interpretations of them. The acknowledgement of feedback into the central dogma—the idea that proteins or other chemicals can influence at least the rate of flow in genetic information—did not come until the proposal of the 'operon theory of gene control' in 1961. François Jacob and Jacques Monod put forward the operon theory following the discovery that in bacteria the expression of genetic information is controlled in part by the binding of specific regulatory proteins to control sequences situated at the beginnings of genes (see Monod 1971).

Control by professional groups of the legitimization of knowledge and the dissemination of relevant information is important in determining what is accepted as 'valid' knowledge. Jacob's and Monod's research material and methodology were better understood by most molecular biologists than was McClintock's work with maize. In contrast to the response by the scientific community to McClintock's work, the operon theory was quickly assimilated into the body of knowledge of molecular genetics and marked the entry of biochemical genetics into molecular biology (Keller 1983).

The operon model postulates that in the bacterial genome there exist genetic elements which control the expression of genes in their locale. In this respect the operon model is compatible with McClintock's pre-molecular explanation of the mechanisms of gene control which she had observed operating in maize. This similarity was intially overlooked. However, once it was acknowledged by Jacob and Monod, McClintock's work began at long last to be taken seriously by the wider community of molecular biologists. As we shall argue in Chapter 4, the insights into gene control made possible by McClintock's work has had a profound effect on our perception of the dynamics of the cell's ecosystem.

For the Swedish botanist, Carl Linnaeus (1707–78), founder of modern taxonomy, naming things was the very foundation of science. The revolution in evolutionary thought propounded in 1859 in the thesis of Charles Darwin (1809–82) and A.R. Wallace (1823–1913) concerned the role of the phenomenon of variation between individuals within a species. Contrary to the classical typological philosophy, which regarded variation as deviation from the ideal form, the Darwin–Wallace thesis envisaged evolution as the conversion of the random variation within an interbreeding group into variation between groups in space and time.

The acceptance of the isolation of the germ-plasm—a refutation of the theory of inheritance of acquired characteristics and a sentiment perpetuated and echoed more than fifty years later in the central dogma of genetics, a guiding principle in unravelling the genetic code in the founding of molecular and biochemical genetics — marked the adoption of a closed, static approach to heredity, which fixes species as securely as the seventeenth- and eighteenth-century

taxonomical classification of natural theology, and distinguishes twentieth-century classical and molecular genetics
from the open, dynamic, nineteenth-century views on
variation.

The reductionist methodology, which is underpinned by
a world-view emphasizing stability, order, uniformity and
equilibrium, clashes with the evolutionary world-view characterized by process, change and non-equilibrium (see, for
example, Prigogine and Stengers 1984). For most scientists,
however, the indubitably remarkable achievements of
molecular and biochemical geneticists from the 1950s
onwards are seen as a triumph for reductionism.

2   Genetic Engineering

Genetic engineering is the manipulation of heredity or the hereditary material. The aim of genetic engineering is to alter cells and organisms so that they can produce more or different chemicals or perform better or new functions.

People have tampered with heredity for as long as they have cultivated crops and bred livestock and are responsible for countless alterations of the inherited properties of life forms on the planet. The ancient craft of genetic engineering was mediated through artificial selection and hybridization.

Artifical selection engineers the inherited properties of future generations by carefully choosing their progenitors. Those subjects with desired traits are selected for breeding and those considered undesirable are excluded from the breeding population. The tangible small-scale evidence of inherited change engineered by breeders and farmers through artificial selection among naturally occurring variants were proclaimed by Charles Darwin to be evidence for the theory of evolution by natural selection. Hybridization is a genetic engineering technique used to broaden the range of genetic traits from which to select those deemed desirable. Traditional hybridization, in which the inherited traits of different species are combined by interbreeding, was a well-established practice by the eighteenth-century.

Genetics is a twentieth-century science which has transformed the ancient craft of genetic engineering into a modern science-based technology. The basic principles of classical genetics, which were developed between 1910 and 1940, changed the study of inheritance from a descriptive, anecdotal account of the outcome of various hybrid crosses, to a rigorous science (Magner 1972).

In the early part of the twentieth century, classical geneticists developed artificial mutagenesis—a method of generating new inherited traits. Artificial mutagenesis induces structural damage to the genetic material of living organisms by exposing them to chemical and physical mutagenic agents, such as X-rays and ultraviolet light. Artificial *in vivo* mutagenesis—mutation of living organisms—is not a precise technique and the outcome in terms of specific traits cannot be predicted. In general, *in vivo* mutagenesis generates strains which have lost characteristics rather than strains which have acquired a new function or property.

Application of the principles of classical genetics had a profound effect on crop plant and domestic animal breeding. Rule-of-thumb breeding procedures were replaced with rational regimes of artificial selection and hybridization, and artificial *in vivo* mutagenesis was used to generate genetic variation.

Scientific breakthroughs in microbiology, biochemistry and molecular biology since the Second World War have created a molecular revolution in genetics which has spawned two fundamental innovations for genetic engineering—recombinant DNA technology and hybridoma technology.

Recombinant DNA technology was generated by the pace and magnitude of the research effort into the functioning of DNA. In conjunction with other techniques for manipulating DNA, recombinant DNA technology has endowed the genetic engineer with a whole new order of techniques for genetic manipulation—*micro*genetic engineering.

Microgenetic engineering enables the genetic engineer to decode, compare, construct, mutate, excise, join, transfer and clone specific sequences of DNA and hence manipulate the inherited characteristics of cells and organisms through precise modification of the hereditary material itself at a molecular level.

Hybridoma technology originated from tissue culture methods developed for research into the genetic basis of the diversity of antibodies. Antibodies are protein molecules formed within the body of an animal in order to neutralize the effect of a foreign invading protein called an antigen. The product of hybridoma technology is a clone of one specific antibody—monoclonal antibodies.

Recombinant DNA technology and hybridoma technology empower humans with their greatest ever power over the inherited traits of life on the planet and together these two revolutionary new technologies created the basis for the renaissance of genetic engineering.

The new techniques of genetic engineering, which are the product of the molecular revolution in biology, are having significant impacts both in basic knowledge and in applied areas such as agriculture, animal breeding and medicine. They will also vastly extend the range and efficiency of potential biotechnological products and services with which they interface in the design and construction of improved and novel biological agents.

2.1 HYBRIDOMA TECHNOLOGY

Hybridoma technology is a specific kind of cell fusion. Cell fusion, in which the entire contents of two or more cells are fused into a single cell, is an innovation in hybridization developed in the 1960s. It can be used to produce new cell types and new types of organism.

In 1975 Georges Kohler and Cesar Milstein, working at the Laboratory of Molecular Biology at Cambridge University, made a breakthrough in animal cell fusion when they developed a technique for fusing antibody-producing cells from mice immunized with a particular antigen with mouse myeloma cells. This breakthrough, called hybridoma technology, was a fundamental innovation in genetic engineering, the products of which are hybrid cells that produce the versatile monoclonal antibodies.

Myelomas are tumours of the immune system. Myeloma cells are 'immortal', which means they can be cultured in the laboratory indefinitely. Antibodies are part of an animal's immune response system. They are protein molecules which detect the presence of foreign agents, such as chemicals, bacteria and viruses, in the body. Each different antibody is sensitive to the presence of a specific part of a larger-sized molecule or a particular small molecule, which is called an antigen. Each type of foreign agent has a characteristic set of antigens on their surface which enables antibodies to identify them as an invading agent which is foreign to the animal's body. When an antibody-producing cell encounters and recognizes an antigen for the first time, the genetic programme of the cell becomes altered so that it produces the appropriate antibody specific for that antigen for the rest of its life. Cell hybrids of myeloma cells and antibody-producing cells are called hybrid myeloma cells or hybridomas (Fig. 2.1.1) They proliferate in culture

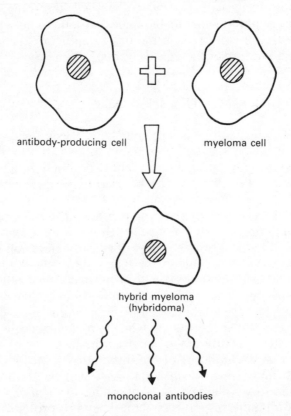

antibody-producing cell   myeloma cell

hybrid myeloma
(hybridoma)

monoclonal antibodies

Figure 2.1.1 Cell fusion to make hybridomas

like cancer cells and produce a clone of one specific antibody—monoclonal antibodies.

One of the main markets for monoclonal antibodies is medical diagnostics. In the field of medical diagnostics about one-sixth of the total market is comprised of immunodiagnostics, which are products based on antibodies, because they are very sensitive and specific and easily 'labelled', for example with radioisotopes, fluorescent dyes and magnetic or colour-coded microspheres, which makes them easy to trace. Antibodies were thought to be too complex to be synthesized chemically and attempts to culture antibody-producing cells *in vitro* have been unsuccessful since such cells neither survive long enough nor produce enough antibodies in culture to become worthwhile sources of antibodies. Furthermore, such systems normally produce mixtures of different antibodies. Using hybridoma technology, however, it is possible to generate clones of antibodies specific to a wide variety of molecules and cell types including proteins, cancer cells, bacteria, fungi and viruses. Monoclonal antibodies are highly specific and pure and have applications in any process which requires the labelling of particular molecules. They have potential medical applications in tissue-typing for grafts and transplants. It is anticipated that by the 1990s there will be an entirely new spectrum of antibody-based diagnostics allowing the rapid, precise and semi-automated diagnosis of ovulation and pregnancy and many infectious diseases, including hepatitis, gonorrhoea, chlamydia, herpes virus and Acquired immune deficiency syndrome (AIDS).

Monoclonal antibodies also have potential medical applications in the treatment of cancer. Cancer chemotherapy involves the use of drugs called cytotoxins which kill cells. Unfortunately, they have widespread side-effects because they damage normal cells as well as killing cancer cells. If cytotoxic drugs were attached to monoclonal antibodies interacting only with cancer cells it is hoped that a 'magic bullet' will be created to target the drug specifically to cancer cells and leave normal cells unharmed. Monoclonal antibodies by themselves have shown some success in treating both leucaemia and colorectal carcinoma (cancer of the large intestine) in mice.

Monoclonal antibodies may be used to neutralize the effects of poisonous substances in the body, such as the toxins produced in disorders like diphtheria and tetanus, food poisoning or drug overdoses. Monoclonal antibodies are believed to be one of the most promising approaches to the treatment and prevention of acquired immune deficiency syndrome (AIDS) (see Kingman 1987).

Monoclonal antibodies specific for antigens on human white blood cells, called leucocytes, are being used in genetic screening tests to predict the susceptibility of different individuals to certain disorders. These human leucocyte antigens (HLAs) are the products of a set of human genes called the HLA system, and studies have revealed that certain HLAs are associated with certain human disorders; for example, the HLA called B27 (HLA-B27) is found in individuals suffering from ankylosing spondylitis, an arthritic condition of the spine (Harsanyi and Hutton 1983).

In the above applications, hybridoma technology is used to engineer novel hybridomas which are cultured in biotechnology reaction vessels to produce monoclonal antibodies as biotechnology product innovations. Hybridoma technology also provides process innovations for biotechnology. On account of their uniform specificity, monoclonal antibodies can be to used label and identify desired product molecules, perhaps producing a gentle method of separation in purifying and recovering product molecules and cells from other substances in the biotechnology reaction vessel.

Hybridoma technology can provide novel enzymes for use in biotechnological processes (Massey 1987). Enzymes, so named from the Greek word *zymos*, meaning 'of yeast', are biological catalysts which bring about the biological processes upon which biotechnology depends. They are naturally-occurring in every living cell and without them the metabolic reactions essential for life, from the digestion of food to the penetration of egg-cells by sperm, would not occur. Monoclonal antibodies and enzymes are both proteins. Hybridoma technology can be used to construct hybridoma cells which produce monoclonal antibodies that behave like enzymes. All that is required to engineer a new enzyme using hybridoma technology is a supply of the molecule with which it is to interact. This molecule becomes a template which stimulates antibody-producing cells to become genetically programmed to make a clone of enzymatic antibodies. These new enzymes are called abzymes because they are a cross between enzymes and antibodies. Thus using hybridoma technology it is possible to create new enzymes to catalyse chemical reactions for biotechnology without understanding the scientific principles of enzyme structure or function.

Naturally-occurring enzymes can only catalyse a limited range of reactions. Since animals are capable of making millions of different antibodies, an almost infinite variety of abzymes can potentially be made, and it is believed that using abzymes it will be possible to create a specific enzyme able to catalyse virtually any reaction. Abzymes have already been produced which can speed up reactions by a factor of 15,000. It is claimed that the use of abzymes could remove the need for high-temperature, high-pressure reactors in the chemical industry, enhancing the conversion from a chemical to a biological mode of production.

2.2 RECOMBINANT DNA TECHNOLOGY

An understanding of the interactions of the components of cells has been an objective of research in molecular biology and biochemistry from the 1950s onwards. In pursuit of this objective, operational systems of cellular processes have been re-created *in vitro* using components of living cells, including molecules of DNA, RNA and various proteins, isolated intact and in retention of their function.

In vitro (L. *vitrum*: glass) is a term used to describe a process which occurs outside of a living organism, usually in a glass vessel in a laboratory. It contrasts with the term *in vivo* (L. *vivo*: live) which denotes a process occurring inside a living organism. By 1960, both DNA and RNA had been successfully synthesized *in vitro* in highly purified cell-free extracts.

The major breakthrough in microgenetic engineering occurred in the early 1970s when genetic recombination was made to occur *in vitro* for the first time. *In vivo* genetic recombination, in which large sections of DNA can exchange places with one another, is not wholly understood. It occurs as part of the normal process of sexual reproduction and is brought about by the cleavage and rejoining of DNA molecules. It is through the process of *in vivo* genetic recombination in the formation of sex cells that paternally- and maternally-derived chromosome pairs break and rejoin with what is normally a reciprical exchange of chromosome parts, with the result that sex cells receive chromosomes of novel genetic combinations.

In vitro genetic recombination is molecular hybridization. It is the precise excision and joining of DNA fragments on the laboratory bench by exploiting enzymes isolated from bacterial cells together with the inherent pairing affinity of the duplex DNA molecule.

The isolation of the bacterial enzymes used in *in vitro* genetic recombination which can cleave and splice DNA from different species was a milestone in the development of microgenetic engineering. These enzymes are the various restriction endonucleases, often referred to as restriction enzymes, and DNA ligase. Restriction enzymes are the DNA 'scissors' and DNA ligase is the DNA 'glue'.

Over 300 restriction enzymes have now been identified, each of which is an isolate of a living bacterial cell. Each different restriction enzyme is used like a pair of molecular scissors to cleave DNA molecules at a specific site, the restriction site, which is characterized by a short sequence of four to ten bases. Their nomenclature is derived from the organism from which the enzyme is isolated, thus *Eco* RI is the name of the first, I, restriction enzyme, R, to be isolated from the bacterium *Escherichia coli (E. coli), Eco*.

Figure 2.2.1 shows the restriction site, a sequence of six base pairs, recognized by the restriction enzyme *Eco* RI. When a piece of DNA with the *Eco* RI restriction site is exposed to *Eco* RI, the enzyme will cut each strand of the DNA in a specific place, the cleavage site, forming what are known as *Eco* RI restriction fragments, as shown in

```
-X-X-X-X-X-X-G-A-A-T-T-C-X-X-X-X-X-X-
-X-X-X-X-X-X-C-T-T-A-A-G-X-X-X-X-X-X-
```

G denotes the nucleotide base guanine
A denotes the nucleotide base adenine
T denotes the nucleotide base thymine
C denotes the nucleotide base cytosine
X denotes any of the above nucleotide bases, paired in accordance
   with the base-pairing rule

Figure 2.2.1 The restriction site of the restriction enzyme ECO RI

```
-X-X-X-X-X-X-G-          -A-A-T-T-C-X-X-X-X-X-X
-X-X-X-X-X-X-C-T-T-A-A-   -G-X-X-X-X-X-X-
```

G denotes the nucleotide base guanine
A denotes the nucleotide base adenine
T denotes the nucleotide base thymine
C denotes the nucleotide base cytosine
X denotes any of the above nucleotide bases, paired in accordance
   with the base-pairing rule

Figure 2.2.2 ECO RI restriction fragments

Figure 2.2.2. Provided flanking restriction sites can be identified, the judicious choice of restriction enzymes enables the microgenetic engineer to excise specific genes from a DNA sample.

The next stage of *in vitro* genetic recombination utilizes the natural affinity between single-stranded DNA sequences which are complementary to each other. Figure 2.2.2 shows that the restriction enzyme *Eco* RI does not cut the two strands of the DNA molecule symmetrically; a cleavage site on one strand is not in line with that on the other. Such cleavage leaves single-stranded DNA sequences on each end of a restriction fragment. These single-stranded end sequences are known as sticky ends. Regardless of the source of DNA, all *Eco* RI restriction fragments will have the same sticky ends, –T–T–A–A– on one end and –A–A–T–T– on the other, which are complementary to each other. All restriction fragments made using the same restriction enzyme will have the same sticky ends.

At low temperatures, single-stranded DNA fragments which have complementary base sequences will pair together, A pairing with T, and C pairing with G, along their length to produce double-stranded DNA. DNA fragments between which *in vitro* genetic recombination is desired are given complementary sticky ends so that at low temperatures the fragments will form associations through complementary base-pairing.

The final stage of *in vitro* genetic recombination relies upon the natural binding properties of the molecular 'glue', DNA ligase. DNA ligase is an enzyme whose name derives from the Latin word *ligare* meaning to bind. It catalyses the joining together of DNA fragments and is used in *in vitro* genetic recombination to reinforce the associations formed between the DNA single-stranded 'sticky ends'.

Using *in vitro* genetic recombination, DNA molecules of diverse origin can be spliced together in novel combinations. The resultant hybrid DNA molecule, called recombinant DNA, can be from two different species or a mixture of natural and synthetic DNA.

Restriction enzymes enable genetic engineers to cleave DNA molecules at specific points to produce discrete reproducible fragments having unique sequences. The discovery of this novel facility soon led to the development of powerful techniques for rapidly sequencing DNA so that the genetic message of specific DNA fragments could be decoded. This new capability was shortly followed by the emergence of convenient methods for synthesizing moderately long pieces of DNA with defined base sequences (Watson *et al.*, 1983). The ability to decode and synthesize DNA molecules *in vitro* endowed the genetic engineer with the power to 'read and write' in the language of genes. Through the combination of *in vitro* genetic recombination and the techniques of synthesizing DNA molecules *in vitro*, microgenetic engineering can redesign genes and their products *in vitro* by site-specific mutagenesis. Unlike *in vivo* mutagenesis, where there is limited control over the nature and location of the mutations induced, *in vitro* site-specific mutagenesis is the creation of predetermined mutations at predetermined sites in a DNA molecule. Using *in vitro* site-specific mutagenesis, fragments of DNA can be deleted at specific sites in a DNA molecule. Similarly, sequences of bases, which can be novel constructs designed in the laboratory or which have been extracted from other genes, can be added to or substituted for existing base sequences.

In vitro genetic recombination is a laboratory bench exercise, the product of which is an engineered DNA molecule called recombinant DNA. The potential of *in vitro* genetic recombination to genetically engineer living cells and organisms can only be realized by inserting recombinant DNA into living target host cells. Recombinant DNA technology is the name given to the combination of *in vitro* genetic recombination techniques with techniques for the insertion, replication and expression of recombinant DNA inside living cells. It has the potential to combine inside living organisms the genetic characteristics of different species with fewer of the strictures of hybridization. For example, bacteria can be endowed with selected inherited traits of humans, and plants can be improved using genes from bacteria. In contrast with cell fusion, which is more useful for combining large parts of the genome, especially where the characteristics of interest are controlled in a complex manner by a large number of genes, recombinant DNA technology is useful when small numbers of individual genes controlling known gene products are involved.

One method of inserting foreign recombinant genetic material into living cells is micro-injection, a technique which pre-dates the techniques for *in vitro* genetic recombination, although it is now used in conjunction with them. In micro-injection the foreign gene, possibly attached to gene control sequences to increase and regulate the rate of its expression, is injected into the host cell.

Another method of inserting foreign genes into host cells is to use self-replicating carrier molecules called vectors. A vector is a self-replicating entity which is used as a 'vehicle' to transfer foreign genes into living cells and then replicate and possibly also express them. Two broad classes of vectors have been developed for recombinant DNA technology; these are plasmids and viruses.

Plasmids are circular double-stranded DNA molecules which are usually found in procaryotes, for example, in bacteria. For most of the time they replicate autonomously, independently of the DNA molecule which constitutes the bacterial genome. The type of plasmids employed as vectors in recombinant DNA technology are 'relaxed-control' plasmids, which replicate inside their bacterial host to produce between ten and several hundred copies of themselves. Although some plasmids carry no coding information at all, others carry a few genes whose products are of benefit to their bacterial host under abnormal culture conditions. Examples are genes whose products confer on the bacterial host resistance to a given antibiotic, or genes that code for enzymes that liberate energy from compounds not commonly available in the environment, such as camphor or salicylic acid (Day 1982; Watson *et al.* 1983).

It is becoming increasingly clear that many micro-organisms which show pathogenic properties in humans and in plants only do so because they carry a particular plasmid. An example of such a plasmid-bacterial association is the Ti plasmid and the bacterium called *Agrobacterium tumefaciens* (*A. tumefaciens*). *A. tumefaciens* infects higher plants, in some of which it causes a cancerous growth called a crown gall tumour. It induces crown gall tumours by using Ti plasmids to inject genetic material which integrates into the genome of cells of the infected plant where it is expressed as a protein. Genetic engineers place much hope on Ti plasmids as vectors for the introduction and expression of foreign genes in plant cells. By manipulating the natural integrative properties of Ti plasmids genetic engineering firms hope that it will be possible to synthesize commercially valuable proteins inside engineered plant cells cultured on a large-scale.

Viruses consist of a nucleic acid molecule, either DNA or RNA, encased in a protein coat and, in some instances, further enclosed in an outer membrane. Viruses are capable of reproduction only when inside a living host cell and are thus termed 'obligate intracellular parasites'. A virus infects a host cell by attaching to its surface and injecting its genome, embodied in its nucleic acid molecule, into the cell leaving the viral coat on the outside. Following infection, the viral genome enters either the lytic phase of its life-cycle, in which multiple copies of the virus proliferate in the host cell, or the dormant lysogenic phase.

In the lytic phase the viral genome arrogates the replicative apparatus of the host cell and causes it to make a large number of copies of the viral genome and to translate viral genes into viral proteins, which include enzymes and proteins for the viral coat. Viral enzymes then catalyse the packaging of individual copies of the viral genome into viral protein coats and the newly assembled viruses escape from the cell to carry the infection elsewhere by 'horizontal' infection of new cells.

Alternatively, the virus may enter the dormant lysogenic phase and be transmitted from cell to cell 'vertically' through the host cell-line. In this quiescent phase, a double-stranded DNA version of the viral genome, called the provirus, integrates into the genome of the host. Unless it is recognized as 'foreign' DNA and destroyed by the host cell, or activated, the viral genome remains integrated latently in this temperate state within the host genome, where it is replicated with the host DNA and thus passed 'vertically' from generation to generation through the host cell-line.

At some later stage an integrated provirus may become activated, usually as a result of some damage to the host chromosome, and change to the lytic phase of the life-cycle. The provirus is excised from the host genome and gives rise to the generation of a clone of new viruses which burst out of the cell.

When vectors are used in recombinant DNA technology, the genetic molecule of the vector, be it plasmid or virus, is cleaved and the gene to be transferred is spliced in, creating what is called a recombinant vector, which bears the foreign gene recombined within its genome. Host cells are then infected by recombinant vectors which carry the foreign gene into them. The vectors replicate inside the host cell to produce multiple copies of themselves, thereby producing clones of the gene inserts. Expression vectors are vectors which are especially constructed to increase the efficiency with which foreign genes are expressed as messenger RNA (mRNA) and protein inside host cells.

The pioneering gene transfer experiments using plasmid vectors were performed in 1973 and 1974. In the first experiment two different antibiotic-resistance plasmids of the bacterium, *E. coli* were cleaved and then recombined together to form a hybrid plasmid bearing genes for resistance to two different antibiotics, tetracycline and kanamycin (S. Cohen *et al.*, 1973). Figure 2.2.3 is a simple schematic representation of this first pioneering experiment.

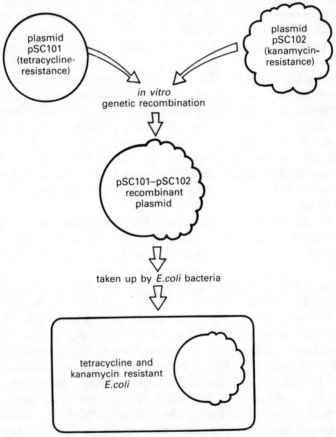

Figure 2.2.3 The first recombinant DNA experiment carried out by Cohen and Boyer et al.

The two antibiotic-resistance plasmids, called pSC101 and pSC102, were selected from a collection of plasmids housed at Stanford University. The nomenclature is 'p' for plasmid, 'SC' for Stanley Cohen, and '101' for the one-hundred-and-first plasmid they tested for its suitability for this experiment. Each of the two plasmids is a circular DNA molecule containing one recognition site for the restriction enzyme *Eco* RI which is external to the gene which provides resistance to an antibiotic: tetracycline in the case of pSC101 and kanamycin in the case of pSC102. The antibiotic-resistance genes were used as markers to identify the recombinant clones.

The circular DNA molecules of both types of plasmid were cleaved using the restriction enzyme *Eco* RI, producing linear plasmid DNA molecules with sticky ends. Cleaved plasmid DNA molecules of both types were incubated together to enable *in vitro* recombination to occur. The resultant recombinant plasmids were of three types: pSC101–pSC101; pSC102–pSC102; and the hybrid recombinant pSC101–pSC102.

Recombinant plasmids of all three types were then incubated with *E. coli* in calcium chloride solution. Calcium chloride solution makes bacterial cell walls 'leaky' and impervious to transformation by surrounding naked DNA molecules. When relaxed-control recombinant plasmids are added to plasmid-free bacteria in calcium chloride solution, they are taken up to yield bacteria that will soon contain multiple copies of recombinant plasmids.

Bacteria which had taken up the hybrid pSC101–pSC102 plasmid were screened for using a medium that contained both the antibiotics, tetracycline and kanamycin; only bacteria with recombinant plasmids bearing tetracycline-resistance from pSC101 and kanamycin-resistance from pSC102 would be able survive on such a medium. Some of the *E. coli* cells that were infected with recombinant plasmids grew into colonies of *E. coli* that were resistant to both of the antibiotics, demonstrating that genes could be transferred into, cloned and expressed in bacteria using transformation with recombinant plasmids.

In the second pioneering experiment the technique of *in*

vitro genetic recombination was repeated using a penicillin-resistance plasmid called p1258 from a species of bacterium called *Staphylococcus aureus* (*S. aureus*) recombined with pSC101 plasmids of the bacterium, *E. coli* (A. Chang and Cohen 1974). Some *E.coli* were transformed by the hybrid recombinant plasmid and inside such bacteria the penicillin-resistance gene from *S. aureus* plasmid p1258 was replicated and expressed. Plasmid p1258 is unable to replicate in *E. coli* on its own. This experiment demonstrated that recombinant DNA techniques make it possible for plasmid genes from one species of bacterium to be transferred into, replicated and expressed in another species of bacterium.

The third pioneering experiment involved the transfer of a gene from a eucaryotic organism, the African clawed toad, *Xenopus laevis*, into *E. coli* bacteria (Morrow *et al.* 1974). Once again pSC101 was employed as a cloning vector. *Eco* RI and DNA ligase were used to splice pSC101 with toad DNA coding for a cellular component called a ribosome. Colonies of bacteria descended from *E. coli* transformed by the hybrid recombinant plasmid were found to be expressing a gene that hitherto had only been expressed in toads. This experiment demonstrated the successful insertion, cloning and expression of a gene from a higher animal in bacteria.

Once the feasibility of using plasmids as vectors in recombinant DNA technology had been demonstrated, more plasmid vectors were developed. Nowadays, plasmid vectors constructed for gene cloning and gene expression applications in research laboratories and in industrial processes are amongst the commercial products of the genetic engineering industry.

To date almost all recombinant DNA work has been accomplished with *E. coli* as the host cell and a range of vectors for use with *E. coli* is available. There are fewer vectors for inserting, cloning and expressing foreign DNA in the cells of bacteria other than *E. coli* or in plant and animal cells, although such vectors are under development. So-called 'shuttle vectors', for example, are vectors which contain genetic sequences that enable them to clone and express foreign genes in both bacterial and eucaryotic cells.

Viruses infect the cells of plants, animals and bacteria and are regarded as more efficient vectors than plasmids for gene insertion into certain cells. Tobacco mosaic virus is currently under investigation as a vector to introduce new genes into tobacco plants. Simian virus (SV40) and human retroviruses are animal viruses which, it is suggested, will prove useful as vectors in the transfer and expression of foreign genes in mammalian cells. In 1987, a new type of cloning vector for cloning large DNA fragments in yeast cells was constructed. Using these new vectors, called yeast artificial chromosomes (YACs), it is possible to clone DNA fragments an order of magnitude larger than was previously possible (Nasmyth and Sulston 1987).

Recombinant DNA technology, it is widely claimed, places the genetic resources of the biosphere at the disposal of the genetic engineer. The potential of this new technology is that it can take these genetic resources, tailor them using site-specific mutagenesis and then use them to design and construct molecules, microbes, cells and organisms to suite commercial requirements.

Like hybridoma technology, recombinant DNA technology is a source of both product and process innovations for the biotechnology industry. Biotechnology processes materials through the exploitation of biological agents on a large-scale. Prior to the advent of recombinant DNA technology and cell fusion technology, the range of products and services provided by biotechnological processes was largely confined to the natural enzymatic capabilities of those biological agents amenable to large-scale culture, principally certain micro-organisms.

Many types of animal cells are not amenable to large-scale culture and this has been a limitation to their exploitation as a source of animal cell products in biotechnology. One approach to this limitation has been the development of techniques for immobilizing animal cells so that they can be cultured (Feder and Tolbert 1983; Spier 1982). Hybridoma technology and recombinant DNA technology are also approaches to this problem, which in different ways have extended the range of products and services which biotechnology can provide to include the products of animal cells. Hybridoma technology permits the synthetic capability of antibody-producing cells to be tapped by fusing them with malignant cells which can be cultured. Recombinant DNA technology permits the genetic resources of cells which are not amenable to large-scale culture to be tapped by inserting their genes into cells or organisms which can be cultured on a large-scale.

The first human gene to be expressed in bacteria was the gene for somatostatin, a small hormone that controls the release of other hormones from the pituitary gland. This was accomplished in 1977 by scientists at Genentech in the USA, the company which synthesized the hormone insulin in microbes in 1978, and human growth hormone in 1979. Scientists at the Swiss genetic engineering company, Biogen, engineered microbes to synthesize human alpha-interferon in 1980. The importance of the microbial cloning and expression of the alpha-interferon gene for the future of recombinant DNA technology was that it demonstrated that it was possible to use recombinant DNA techniques to clone and express a gene about the structure of which nothing was known, and which produces a very small amount of product.

Once it had been demonstrated that using recombinant DNA techniques it is possible to insert, clone and express human DNA inside microbial cells, companies began to search for other suitable human protein molecules to synthesize in microbes. There are believed to be over 50,000 different proteins in the human body (excluding antibodies) of which very few are currently in use in medicine, indeed fewer that 2 per cent have been identified. Human proteins which have been identified as targets for microbial synthesis using recombinant DNA technology include hormones and drugs, such as interferon, which are too complex to be made economically by traditional chemical methods; blood-clotting factors as alternatives to extracts from blood for use in the treatment of haemophiliacs; anti-clotting agents to stimulate the breakdown of blood clots which may precede strokes; endorphins for the treatment of addiction to morphine drugs and depression and possibly for pain relief; cell membrane constituents; and nerve cell growth stimulators.

As with hybridoma technology, a potential diffusion of recombinant DNA technology into biotechnology is in enzyme engineering. Recombinant DNA has been used to convert micro-organisms into super-producers of commercially valuable enzymes which are purer than the same enzymes extracted from animal or plant sources. *In vitro* site-specific mutagenesis may be used to modify microbially cloned enzymes by modifying the genes that encode them. In this way it is envisaged that recombinant DNA technology will furnish biotechnology with enzymes that are tailor-made for specific purposes in manufacturing processes.

Recombinant DNA technology permits the genetic constitution of microbes used in biotechnology to be improved in predetermined ways. Their synthetic, degradative and leaching properties can be enhanced and the range of substrates that they utilize can be altered; they can also be modified to withstand adverse culture conditions and their manufacturing efficiency can be increased. However, there are problems with the expression of certain animal and plant genes in bacterial cells. In recent times there has been a shift by genetic engineering companies away from microbial cloning towards the use of animal and plant cells and even whole animals as hosts for cloning and expressing valuable foreign genes. Theoretically, if the appropriate gene is inserted and expressed, any protein molecule can be synthesized in this way. Moreover, the range of target molecules is not restricted to protein molecules. Just as yeast cells are used in brewing and fermentation to produce alcohol, so the biosynthetic pathways of cells, microbes and other organisms can be genetically manipulated so that they combine and break down molecules in their immediate environment, thereby producing marketable non-protein molecules.

The range of product molecules which can potentially be produced by recombinant cells or organisms encompasses virtually any organic molecule. Organic molecules are molecules which are produced by living organisms. They include products currently made from petrochemicals and extracted from natural sources such as plastics, flavour and perfume materials, synthetic rubber, pharmaceuticals including vaccines, pesticides, industrial solvents, oils and lubricants.

Initially, the products of recombinant cells and micro-organisms have been process innovations as the new techniques of microgenetic manipulation have been applied to produce existing products by novel methods. In the next phase of its diffusion into industry, recombinant DNA technology will lead to a spectrum of product innovations as the techniques of microgenetic manipulation are used to design new products and engineer living cells and organisms to synthesize them.

2.3   THE GENETIC ENGINEERING INDUSTRY

The new techniques of genetic engineering—microgenetic engineering and hybridoma technology—are powerful tools for research and industrial applications. Genetic engineering is being used in human health care research to increase our knowledge of the molecular genetics of immunological processes, development, ageing and cancer. It is used directly in human health care for the identification of genes responsible for birth defects for genetic screening. A long-term objective of genetic engineering in human health care is the correction of single-gene disorders through the induction of the genes for missing enzymes and gene-replacement therapy. There are even projects underway to map and determine the base sequence of the entire human genome (Joyce 1987; Barinaga 1987b).

A new industry—the genetic engineering industry—based upon these revolutionary new technologies, has already emerged. The genetic engineering industry develops, manu-factures and sells genetically manipulated organisms, cells, enzymes, vectors and genes which produce goods and perform services in scientific research, health care, the chemicals industry, the food industry, agriculture, energy, environmental control and resource recovery.

Table 2.3.1: *Diffusion of the new techniques of genetic engineering to the year 2000*

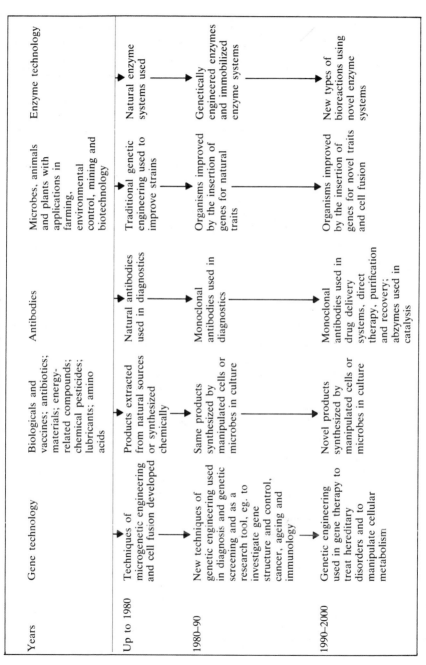

Years	Gene technology	Biologicals and vaccines; antibiotics; materials; energy-related compounds; chemical pesticides; lubricants; amino acids	Antibodies	Microbes, animals and plants with applications in farming, environmental control, mining and biotechnology	Enzyme technology
Up to 1980	Techniques of microgenetic engineering and cell fusion developed	Products extracted from natural sources or synthesized chemically	Natural antibodies used in diagnostics	Traditional genetic engineering used to improve strains	Natural enzyme systems used
1980–90	New techniques of genetic engineering used in diagnosis and genetic screening and as a research tool, eg. to investigate gene structure and control, cancer, ageing and immunology	Same products synthesized by manipulated cells or microbes in culture	Monoclonal antibodies used in diagnostics	Organisms improved by the insertion of genes for natural traits	Genetically engineered enzymes and immobilized enzyme systems
1990–2000	Genetic engineering used in gene therapy to treat hereditary disorders and to manipulate cellular metabolism	Novel products synthesized by manipulated cells or microbes in culture	Monoclonal antibodies used in drug delivery systems, direct therapy, purification and recovery; abzymes used in catalysis	Organisms improved by the insertion of genes for novel traits and cell fusion	New types of bioreactions using novel enzyme systems

Since the end of the 1970s there has been a marked growth in the genetic engineering industry. The anticipated diffusion over time of the product and process innovations of this new genetic engineering industry is presented in Table 2.3.1.

Optimistic market expectations for the products and services of the genetic engineering industry has engendered a cluster of innovations which include bought-in technologies for genetic engineering (see also Wheale and McNally 1986). Amongst these are: computer software for the computer-aided design of genes and their products and for DNA sequence analysis; cell culture systems; automated DNA synthesizers and decoders which facilitate the bio-chemical manipulation of DNA and accelerate iterative processes; and kits and apparatus for routine physical manipulations of genetic material, such as DNA, RNA and plasmid purification, DNA probe labelling, site-specific mutagenesis and DNA cloning. Computerized databanks and networks of DNA and protein sequences, such as the Nucleotide Sequence Data Library of the European Molecular Biology Laboratory (EMBL) in Heidelberg, West Germany and GenBank in the USA have been established to facilitate information access.

The Austrian economist Joseph Schumpeter, considered innovation to be the driving force of economic growth (Schumpeter 1939), though some economists argue that the impetus to economic growth comes not from the first innovations but from a pattern of change associated with a cluster of innovations conducing to the growth of new industries amounting to the emergence of new technological systems (Clark *et al*. 1984). Computerized automation, electronics and laser technology together with diffusion of the new technologies spawned by the molecular revolution in the life sciences represent such a cluster of fundamental innovations. Indeed, many futurologists forecast that the application of recombinant DNA technology and hybridoma technology by the genetic engineering industry will have revolutionary impacts both on the growth of scientific knowledge and on applied areas including agriculture, biotechnology and health care before the end of the century.

3   Guardians or Jackasses?

Recombinant DNA techniques, pioneered by the research teams of Stanley Cohen and Herbert Boyer in the USA in 1973 and 1974, were heralded as a new technological revolution. On the one hand, it was believed that these techniques would provide new social benefits including greater understanding of the causes and treatment of cancer, more effective pharmaceutical products, a second 'green revolution' providing a greater abundance of food crops for less economically developed countries and new approaches to energy production and pollution control. Moreover, the use of recombinant DNA techniques opened up new opportunities for scientific research into the mechanism of gene expression in higher organisms. On the other hand, many scientists were uncertain of the estimated degrees of risk of conducting recombinant DNA experiments and fearful of potential biohazards.

In 1974, in an unprecedented act of 'responsible science', the international scientific community voluntarily accepted a self-imposed moratorium on certain recombinant DNA experiments which they considered to be too risky to be undertaken. However, by 1978 a new consensus had emerged among the scientific community. The new consensus was that the hazards of recombinant DNA work had been greatly overestimated, and that if reasonable laboratory practices were observed, the risks of epidemics and ecological catastrophe were so slight as to be insignificant. 'We were teenage jackasses', is the oft-quoted reply of James Watson when asked in 1979 to explain the apparent *volte*

41

face of molecular biologists over the risks of recombinant DNA techniques. By the end of the decade the scientific community had successfully defused the issue of hazards arising from recombinant DNA work, and James Watson, John Tooze and David Kurtz expressed the relief of most molecular biologists when they wrote in the preface to their book, *Recombinant DNA*, published in 1983, 'Fortunately, our worst fears proved unfounded, and today most forms of recombinant DNA research are no longer subject to any effective form of regulation'. In this chapter, we analyse the development of the recombinant DNA debate and address the processes by which this *volte face* came about.

3.1   REGULATORY POLICY IN THE USA

The two major countries participating in the recombinant DNA debate were the UK and the USA. Public alarm in the USA concerning the risks of genetic engineering focused on the planned gene transfer experiments of Paul Berg's research team at Stanford University announced in 1971. The team planned to study the tumour virus, Simian Virus 40 (SV40), by cloning it in *Escherichia coli*, the common bacteria of the human intestinal tract and sometime human pathogen. SV40 virus was known to be capable of inducing cancer in hamsters. The consequences of the colonization of the intestines of humans by *E. coli* that had been infected with cancer-inducing DNA from SV40 were greatly feared.

The subsequent demonstration by the research teams of Cohen and Boyer of the potential of recombinant DNA techniques to manipulate and combine DNA from widely different species intensified public concern over the attendant biohazards and with the ethical, moral and political questions that the use of these new techniques raised.

Much of the concern about the health hazards of recombinant DNA technology focused on the use of *E. coli* bacteria as host cells for recombinant DNA. *E. coli* are able to infect and colonize humans and are normal inhabitants of the human gut. When *E. coli* are used as hosts for recombinant DNA, humans are at risk of infection by recombinant *E. coli* which contain DNA which is foreign to the bacteria and to humans. The expression of this foreign DNA within the digestive system of humans could be harmful and could transform *E. coli* into human pathogens.

A second concern was the consequences of the use of antibiotic-resistance plasmids, constructed to contain mixtures of many antibiotic-resistance genes, as vectors and markers in gene transfer experiments. Bacteria acquire resistance to antibiotics from antibiotic-resistance plasmids which reside within them. Antibiotic-resistance plasmids carry one or more genes which code for molecules which protect their host bacterium from the harmful effects of specific antibiotics.

Antibiotic-resistance plasmids are especially useful in recombinant DNA technology. First, as vectors, they can be spliced to contain a foreign DNA insert which they can carry and clone in target host bacteria. Second, when antibiotic-resistance plasmids are used as vectors, the antibiotic-resistance genes are useful genetic markers for identifying which bacterial colonies contain the recombinant plasmid vectors. Such colonies might, for example, be distinguishable because they have acquired resistance to certain antibiotics in the growth medium, whereas bacterial colonies which have not taken up recombinant antibiotic-resistance plasmids will not have acquired this resistance.

The widespread regular use of antibiotics in medicine and animal husbandry is a powerful selective force, eliminating those microbial strains which are not protected by antibiotic-resistance plasmids and selecting for those strains which possess such plasmids. The use of the antibiotic penicillin to treat the sexually transmitted disease gonorrhoea among the Allied soldiers during the Italian Campaign in the Second World War, enabling them to be back on duty within forty-eight hours, gave the Allies a military advantage over the Germans who were without antibiotics at that time. However, following the use of antibiotics, by the late 1950s a new and more serious strain of *Neisseria gonorrhoeae*, the bacteria responsible for gonorrhoea, which was resistant to penicillin, began to proliferate. It is also alleged that as a result of massive over-prescription of antibiotics to American troops in the Vietnam War, there have been epidemics of gonorrhoea, salmonella and gastro-enteritis in that geographical area.

Outbreaks of antibiotic-resistant strains of *Staphylococcus* bacteria, which can infect ulcers and sores and prevent surgical wounds from healing, have occurred in hospitals in the UK, France, the USA, South Africa, the Middle East and Australia. Certain strains of microbe have acquired resistance to more than one antibiotic as a result of the transfer of antibiotic-resistance plasmids from one bacterium to another during conjugation (mating) (WHO 1987).

Concern over the use of antibiotic-resistance plasmids as vectors in recombinant DNA work was that, were the host bacterium to escape, its proliferation and spread would be difficult to contain on account of the protection it would have from the antibiotic-resistance genes on the recombinant plasmids. There was also concern that their use would contribute to the further spread of bacterial drug resistance through the passage of the recombinant plasmid from the escaped bacterial population into resident microbial flora in the human gut, which would then become transformed into antibiotic-resistant strains. Similarly, in sewage, pathogenic bacterial strains could acquire protection from the effects of antibiotics by the passage of antibiotic-resistance recombinant plasmids from recombinant bacteria which escape. Figure 3.1.1 is a simple schematic representation of the way in which the use of antibiotic-resistance plasmids as vectors in recombinant DNA work increases the risk that recombinant bacteria will proliferate.

A third cause of concern was the 'alchemical' mixing of the products of evolution through 'shotgunning', a process undertaken by researchers in the analysis of the genome of higher plants and animals. Shotgunning is the cleaving of the entire genome of a species into fragments, which are spliced into vectors and transferred into bacteria. Subsequent cell division by the bacteria produces colonies of bacteria and each bacterial colony represents a clone—multiple identical copies—of a fragment of the genome, which is then available in large amounts for

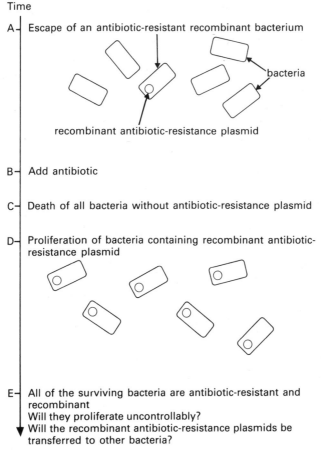

Figure 3.1.1 *The danger of use of antibiotic-resistance plasmids in recombinant DNA work*

further study. The whole set of colonies derived from the shotgunning of an entire genome constitutes what is called a gene library for the species.

Concern was expressed that shotgunning could inadvertently create dangerous new pathogens. The function of approximately 90 per cent of the DNA of most higher organisms has not been characterized and may contain parasites, including viruses, and cancer genes from the species under study. As the name implies, shotgunning is not a precise process. The nature of the foreign gene fragment within a given recombinant bacterial colony derived by shotgunning cannot be predetermined. It was feared that recombinant bacteria produced by shotgunning could be responsible for new diseases of animals and plants, new forms of cancer and novel epidemics. It was also feared that recombinant bacteria would flourish in the biosphere because they are novel life forms against which existing living organisms could not have developed defences and immunities.

The question of the potential hazards of experiments which create recombinant organisms was raised by participants of the 1973 Gordon Research Conference on Nucleic Acids in New Hampshire in the USA. The chairpersons of this conference sent a letter to the president of the US National Academy of Sciences (NAS). This letter, entitled 'Guidelines for DNA Hybrid Molecules', was published in *Science* in 1973 (Singer and Soll 1973).

The response by the NAS was to sponsor the Committee on Recombinant DNA molecules, composed of prominent US scientists and chaired by Paul Berg, to address the risks question. This produced the 'Berg Letter', which called for a partial and temporary voluntary moratorium on experiments involving the introduction of new antibiotic-resistance genes or bacterial toxin genes into bacteria that did not normally carry these genes, and on the introduction of DNA from tumour viruses or other animal viruses into autonomously reproducing DNA elements, such as plasmids. Versions of this letter were published in *Science*, *Nature* and the *Proceedings of the National Academy of Sciences* in July 1974 (Berg *et al.* 1974). In a unique act of 'responsible science' in the history of science, the scientific community worldwide immediately accepted the voluntary moratorium, pending an international scientific meeting to discuss potential biohazards of recombinant DNA work.

In February 1975, 150 participants from all over the world convened at Asilomar in the USA to make recommendations concerning the regulation of applying recombinant DNA techniques. One of the chief reasons for this act of 'self control' was the concern that, as Stanley Cohen of Stanford

University put it, 'If the collected wisdom of this group doesn't result in recommendations, the recommendations may come from other groups less well qualified' (see Wade 1975). Also influencing the decisions of the conferees was the knowledge that their recommendations were going to be scrutinized outside the scientific community as well as within it. With notable exceptions, having drawn up their guidelines, the Asilomar conferees voted to end the voluntary moratorium and to resume recombinant DNA work (Berg *et al.* 1975).

In 1974 the National Institutes of Health (NIH), a federal agency in the USA which funds and regulates federally-funded research proposals, had appointed a new committee, a panel largely composed of molecular biologists, which came to be called the Recombinant DNA Advisory Committee (NIH-RAC). The Asilomar recommendations were submitted in a report to the Assembly of Life Sciences and were then considered by the NIH-RAC, which urged a tightening of the guidelines in a way that would effectively preclude the use of recombinant DNA procedures to study cancer viruses. The NIH-RAC recommendations were approved and subsequently established the pattern for the first NIH Guidelines for Research Involving Recombinant DNA Molecules, issued in 1976 (*Guidelines* 1976).

Under the NIH Guidelines of 1976 each of the various sorts of experiment employing recombinant DNA techniques was classified according to an estimate of the level of risk of potential biohazard it constituted. In addition, certain classes of work were considered to pose a potential hazard so great that they were prohibited. For example, those involving the use of known pathogens or tumour viruses, those involving the introduction of antibiotic-resistance genes into organisms that did not normally carry such genes, those requiring the deliberate release of recombinant organisms and those that required the use of the techniques on a large-scale. For the remaining permitted classes of experiments, following the Asilomar Conference recommendations, the NIH Guidelines defined combinations of physical and biological containment measures to correspond with estimated levels of biohazard, as summarized in Table 3.1.1

Table 3.1.1: *Categories of containment for recombinant DNA experiments (NIH Guidelines, 1976)*

		Biological Containment (for *E. coli* host systems only)		
		EK1	EK2	EK3
Physical containment	P1	DNA from non-pathogenic procaryotes that naturally exchange genes with *E. coli* Plasmid or bacteriophage DNA from host cells that naturally exchange genes with *E. coli*. (If plasmid or bacteriophage genome contains harmful genes or if DNA segment is less than 99 per cent pure and characterized, higher levels of containment are required.)		
	P2	DNA from embryonic or germ-line cells of cold-blooded vertebrates DNA from other cold-blooded animals and lower eucaryotes (except insects maintained in the laboratory for fewer than 10 generations) DNA from plants (except plants containing known pathogens or producing known toxins) DNA from low-risk pathogenic procaryotes that naturally exchange genes with *E. coli*	DNA from non-embryonic cold-blooded vertebrates DNA from moderate-risk pathogenic procaryotes that naturally exchange genes with *E. coli* DNA from non-pathogenic procaryotes that do not naturally exchange genes with *E. coli* DNA from plant viruses Organelle DNA from primates. (For organelle DNA that is less than 99 per cent pure higher levels of containment are required.)	

...that is less than 99 per cent pure higher levels of containment are required.)	Plasmid or bacteriophage DNA from host cells that do not naturally exchange genes with *E. coli*. (if there is a risk that recombinant will increase pathogenicity or ecological potential of host, higher levels of containment are required.)	DNA from non-embryonic primate tissue DNA from animal viruses (if cloned DNA contains harmful genes)
P3 DNA from non-pathogenic procaryotes that do not naturally exchange genes with *E. coli* DNA from plant viruses Plasmid or bacteriophage DNA from host cells that do not naturally exchange genes with *E. coli*. (If there is a risk that recombinant will increase pathogenicity or ecological potential of host, higher levels of containment are required.)	DNA from embryonic primate-tissue or germ-line cells DNA from other mammalian cells DNA from birds DNA from embryonic, non-embryonic or germ-line vertebrate cells (if vertebrate produces a toxin) DNA from moderate-risk pathogenic procaryotes that do not naturally exchange genes with *E. coli* DNA from animal viruses (if cloned DNA does not contain harmful genes)	
P4	DNA from non-embryonic primate tissue DNA from animal viruses (if cloned DNA contains harmful genes)	

Source: C. Grobstein 'The Recombinant-DNA Debate', *Scientific American* 237 (1). 22–33, 1977.

Physical containment measures are designed to prevent or minimize the escape of recombinant organisms. The 1976 NIH Guidelines prescribed four levels of physical containment, designated P1, P2, P3 and P4 (see Table 3.1.1). Physical containment level P1, for what is perceived as the least hazardous classes of experiments, requires little more than good microbiological laboratory practice, whereas a P4 laboratory is a maximum isolation installation, considered appropriate for handling dangerous pathogens.

Biological containment is a form of 'technological fix': a scientific development intended to solve the problem of risk through the use of enfeebled microbial host strains and vectors. Notwithstanding circumvention of the deficiency by mutation or natural (*in vivo*) genetic recombination, the purpose of biological containment is to reduce the potential hazard were the escape of recombinant micro-organisms to occur. Microbial hosts are genetically enfeebled so that they are dependent on laboratory conditions for their survival and are less likely to survive were they to escape into the environment. Vectors are enfeebled so that they are less able to move to a new host strain and are less likely to transfer foreign genes to non-target microbial hosts. The biological containment levels in the 1976 NIH Guidelines are classified from EK1, the lowest level of biological containment, to EK3 (see Table 3.1.1).

Under the 1976 NIH Guidelines, each institution wishing to conduct recombinant DNA work had to appoint a Principal Investigator and form an Institutional Biohazard Committee (IBC). It was the responsibility of the Principal Investigator to apply the NIH-RAC Guidelines in order to assign proposed recombinant DNA experiments in his or her institution to the appropriate containment category. The NIH Guidelines defined the responsibilities of the IBCs as those of advising the institutions on policies, creating a central reference of relevant information, developing a working practices manual for any P4 facilities, and certifying that all safety conditions were met on NIH research applications. Each IBC was required to comprise at least five members drawn from disciplines relevant to recombinant DNA technology, bio-safety and engineering.

Thus, although the major regulatory agency for recombinant DNA work was the NIH, responsibility for monitoring recombinant DNA activities within a given institution and ensuring compliance with the Guidelines was delegated to the Principal Investigator and the IBC.

3.2 REGULATORY POLICY IN THE UK

In the UK public alarm over the escape of pathogenic micro-organisms was triggered by the deaths of two people from smallpox following contamination originating from a laboratory at the London School of Hygiene and Tropical Medicine in 1973. Following the smallpox tragedy, the British Secretary of State for Social Services formed the Working Party on the Laboratory Use of Dangerous Pathogens, chaired by Sir George Godber, the government's chief medical officer, to investigate and report on the safety of laboratories using dangerous pathogens.

The laboratory involved in the smallpox tragedy was not conducting recombinant DNA work, and indeed very little research using recombinant DNA techniques was being pursued in the UK at that time. Whilst the Godber Working Party was deliberating laboratory safety measures, the Berg letter was published in 1974. The UK government responded to the Berg letter by establishing a working party under the auspices of the Advisory Board for the Research Councils (ABRC) to be comprised of scientists and chaired by the distinguished biologist Lord Ashby. The remit of the Ashby Working Party was to make recommendations on the experimental manipulation of the genetic composition of micro-organisms.

The Ashby Report (1975) was produced quickly in order to make it available as a British contribution to the deliberations at the Asilomar Conference in the USA in February 1975. In August 1975, a second UK governmental committee on recombinant DNA work was established by the Department of Education and Science (DES). This second committee, the Working Party on the Practice of

Genetic Manipulation, was chaired by Sir Robert Williams, then director of the British Public Health Laboratory Service. Its remit was to draft a central code of practice and to make recommendations for the establishment of a central advisory service for laboratories using the techniques available for genetic manipulation, and for the provision of necessary training facilities. Whilst awaiting the recommendations of the Ashby Working Party, the Godber Working Party of the Department of Health and Social Security (DHSS), and the Williams Working Party of the DES, the Medical Research Council (MRC) instructed scientists in its laboratories to observe the moratorium on recombinant DNA work proposed in the Berg letter.

The Godber Report of 1975 recommended that a small independent central committee consisting of individuals whose experience would command the confidence of those working in laboratories be established to implement a code of practice for the handling of dangerous pathogens. Acting on the recommendations of the Godber Report, the DHSS established the Dangerous Pathogens Advisory Group (DPAG), a body composed solely of experts on dangerous pathogens. The emphasis of the Ashby Report (1975) was also that safety was best left to those experienced in microbiological techniques. It recommended that each institute conducting recombinant DNA work should have its own biological safety officer to be assisted initially by a national advisory service to advise scientists on appropriate levels of safe laboratory practice. In many respects the Williams Report (1976) enlarged upon the recommendations of the Ashby Report. However, the major difference between the recommendations of Williams and those of Ashby was that the UK national advisory body for genetic manipulations should not be entirely composed of experts in the field. This was in contrast to the NIH-RAC in the USA and the DPAG in the UK. In 1976, as a result of the Williams Report, the Genetic Manipulations Advisory Group (GMAG) was formed by the DES and began work under the chairmanship of Sir Gordon Wolstenholme. Members of the GMAG Committee represented four different categories of interest: eight of the members were scientific or medical experts; four represented employees; two represented management; and four represented the public interest.

The Williams Report differentiated between the requirements for safeguards against the new conjectured hazards of various classes of recombinant DNA experiments and those for the known hazards of handling dangerous pathogens. However, there was a degree of indecision in the UK over whether recombinant DNA research should be treated for the purposes of regulation as being of the same nature as research with dangerous pathogens or whether they should be separated. Medical microbiologists asserted that nothing novel could arise as a result of the use of recombinant DNA techniques that was more hazardous than existing pathogens and that the well-tried practices of medical microbiology would be entirely adequate to contain any potential hazards. Accordingly, the Godber Report recommended that recombinant DNA research should be placed under the auspices of the DPAG. However, it was precisely because of the failure to adhere to the safety practices of medical microbiology that the two smallpox deaths occurred following the contamination at the London School of Hygiene and Tropical Medicine.

The Ashby Report had recommended that recombinant DNA experiments should be limited to genetically enfeebled strains of bacteria. Although relatively greater emphasis was placed on physical containment, the containment safeguards recommended for recombinant DNA experiments in the Williams Report, whose recommendations the GMAG initially acted upon before issuing its first Advisory Notes in 1977, were not substantially different from those issued by the NIH in the USA. In a similar way to the NIH, the GMAG proposed four levels of physical security, Categories I-IV, for laboratories in which work with recombinant micro-organisms was carried out. The NIH Guidelines comprised a comprehensive framework to be interpreted at the laboratory level which assigned classes of recombinant DNA experiments to containment categories on the basis of biological criteria only. The UK system, in contrast, retained the flexibility of a central committee, the

GMAG, which would allocate proposed experiments to appropriate containment categories on a case-by-case basis, taking a broader range of factors, including laboratory facilities and safety record and personnel expertise, into account as well as biological criteria.

Another development in the UK at this time was the establishment in 1974 of the Health and Safety Commission (HSC) under the Health and Safety at Work Act (HSW Act) to monitor health and safety at work. In August 1978 the HSC issued Health and Safety (Genetic Manipulation) Regulations which made it compulsory that notification of all recombinant DNA work in the UK be made to the inspectorate of the HSC—the Health and Safety Executive (HSE)—and to the GMAG.

The GMAG ensured that local biological safety committees were established wherever recombinant DNA research was undertaken. The presence today of a laboratory safety committee in every university or teaching hospital in the UK is in marked contrast to the situation which existed when the GMAG was formed, when hardly a single such establishment had one. As well as exposing the lack of organized and local safety advice, the GMAG revealed the unreliability and defectiveness of some of the physical containment measures which were relied upon and encouraged biological containment through the use of enfeebled hosts and vectors. It persuaded some scientists to do what Ashby called the 'ungrateful science' (research which does not attract accolades) necessary to develop enfeebled strains of bacteria and to carry out experiments to assess the risks of recombinant DNA work (Denselow 1982).

In 1976 the European Molecular Biology Organization (EMBO) standing advisory committee recommended that countries using recombinant DNA techniques should follow either the USA NIH Guidelines or those laid down in the UK Williams Report. The two contrasting approaches to the regulation of recombinant DNA work informed the regulatory framework adopted by most other countries formulating recombinant DNA guidelines.

The institutional responses in the UK and USA to the regulation of recombinant DNA work were largely determined by the contrasting geographical and political structures of the two countries. Although the composition of the GMAG, comprising representatives of industry, employees and the public as well as scientists, constituted a wider representation compared with the NIH-RAC in the USA, members of the GMAG met in camera and were bound by the Official Secrets Act. This is in marked contrast to the open system in the USA, where the deliberations of the NIH-RAC were broadcast under the Freedom of Information Act.

The UK is a unitary state which facilitates a centralized control system in contrast with the federal political system of the USA, with its differential state legislative controls. The UK traditionally favours a case-by-case evaluation of activities which could potentially damage the public welfare, such as those that would create industrial pollution. The USA, on the other hand, has tended to favour formal, relatively inflexible, but decentralized guidelines or legislation to control such activities. These national differences were reflected, by 1976, in the two sets of guidelines for recombinant DNA research: the one created in the USA was, 'encyclopaedic, complex and rigid'; the other, contained in the Williams Report in the UK, was 'flexible, pragmatic and modelled on case-law rather than statute' (Yoxen 1983).

3.3 PUBLIC OPINION

The response to the hazards of recombinant DNA work in the UK was in marked contrast to that in the USA. In the UK, safety procedures were investigated through *ad hoc* governmental-initiated committees, and in contrast with the US public, the British public exhibited little awareness of the issues in the recombinant DNA debate. The more open political system of the USA, together with its greater acceptance of pressure-group lobbying of the government to influence decision-making, fostered an environment which promoted a wider public debate than in the UK, where the deliberations of the working groups were much less public. In the USA, by the beginning of 1976, the policy debate over recombinant DNA research was escalating both in its intensity and in the range of public involvement (Grobstein 1977).

The NIH Guidelines for Research Involving Recombinant DNA Molecules became controversial even before they had been officially released in July 1976. In November 1975 the NIH-RAC invited public comment on its draft recombinant DNA Guidelines. The response from the general public was one of severe apprehension about the development of genetic engineering. The controversy erupted at a hearing of the NIH Guidelines in February 1976 and spread quickly to some of the local communities which were centres of recombinant DNA activity, which included Ann Arbor, Cambridge, Princeton, Berkeley and San Diego.

The NIH Guidelines were criticized by the US public on the grounds that the physical and biological constraints they recommended for recombinant DNA work were inadequate to guarantee effective containment of recombinant bacteria and vectors. Although there was scope in the NIH Guidelines (1976) for the use of other hosts, the preferred enfeebled host was a strain of the common intestinal bacterium *E. coli*, known as *E. coli* K12, which had been weakened by many years of use in molecular genetics laboratories. *E. coli* K12 were credited with being biologically safe because they were not believed to be pathogenic in humans, nor were they considered capable of surviving outside the laboratory or in competition with other microorganisms.

Concern was expressed that the physical and biological constraints of the NIH Guidelines would be inadequate to guarantee effective containment of *E. coli* K12 and that were recombinant *E. coli* K12 bacteria to escape, the recombinant DNA insert might be transferred to other more robust organisms that could survive effectively in humans and in the environment. There was also fear that as a result of genetic manipulation, the weakened *E. coli* K12 strain might be converted inadvertently into a pathogenic strain. Moreover, although most recombinant DNA research employed *E. coli* K12 as hosts, there was scope in the NIH guidelines for the use of other host organisms, of which enfeebled strains had not been developed. The action and function of many genes which might be inserted into these organisms was not well understood.

Another criticism raised in the public debate was the diminished role of the general public in the formulation and administration of the NIH Guidelines. Only a few individual scientists were involved in drafting the early guidelines and in creating the NIH regulatory mechanism. Scientists from other disciplines were excluded and, with the exception of a few lawyers, so were representatives from non-scientific groups. Amongst those who voiced this criticism was Senator Edward Kennedy, who accused molecular biologists of 'making public policy in private' (Gershon 1983).

Under the 1976 Guidelines the major regulatory body for recombinant DNA work was at the NIH. The NIH advisory group responsible for regulating such research, the NIH-RAC, was largely composed of scientists, thus it was alleged that the NIH Guidelines gave recombinant DNA researchers effective self-regulation.

It was also noted during the public debate that there was a potential conflict of interest between the NIH, the largest funder of biomedical research, and its own advisory group, the RAC, the body responsible for regulating such research, particularly as the NIH-RAC is largely composed of scientists with a vested interest in furthering recombinant DNA work.

Another area of concern was the limited control by the NIH over those conducting recombinant DNA work. In the USA the ability of any federal agency to regulate is tied to its substantive grants of statutory authority from Congress. In certain areas the agencies are without regulatory powers until certain requisites are met or established.

The authority of the NIH is derived from the Public Health Service Act. This Act authorizes the Department of Health and Human Services (DHHS) to prevent the introduction, transmission, or spread of communicable diseases. Most of the provisions of the recombinant DNA

Guidelines do not involve preventing the occurrence of communicable diseases. In cases where there is no reasonable basis for believing that the products of recombinant DNA research will cause human disease or a communicable disease, the NIH lacks statutory authority to regulate such recombinant DNA research (Korwek 1981). Therefore, although NIH Rules are regulations that have the force of law, rather than merely 'guidelines' for government-funded institutions and projects, the authority of any US agency to require compliance with the recombinant DNA Guidelines appears to be based upon contract law rather than upon statutory authority to issue substantive rules (Korwek 1980).

In practice, the NIH Guidelines apply to NIH grantees but not to the private sector or even to work supported by other government agencies. Each institution receiving NIH funds for conducting or sponsoring recombinant DNA research became responsible for ensuring that this activity was carried out in conformity with all the safety provisions of the NIH Guidelines; refusal to comply with the Guidelines would result in withdrawal of NIH funding.

Public concern over the potential hazards of recombinant DNA work and dissatisfaction with the content and accountability of the NIH Guidelines and the NIH-RAC, resulted in a public reaction which threatened the future of recombinant DNA research in the USA.

By the spring of 1977 about a dozen bills to regulate recombinant DNA technology had been introduced into Congress. Recommendations submitted to Congress at this time included provisions for external regulation of the field by a presidential commission, federal licensing, complex reporting systems and sanctions which included substantial fines for scientists who transgressed the regulations. Senator Edward Kennedy proposed a bill which advocated public participation in the formulation of science regulation policy (see Gershon 1983). His bill proposed the establishment of an independent national regulatory commission specifically for recombinant DNA research. It was to be comprised primarily of non-scientists, and would provide the minimum regulation required in this field.

The lack of adequate federal regulation resulted in local

initiatives in some areas of the USA. Whilst some individuals and groups requested governmental regulation in the form of legislation, others demanded a complete ban on all recombinant DNA work. In particular, the city of Cambridge in Massachussetts recognized early in the debate that citizens have a right to be involved in decisions about the consequences of any new technology. By July 1976, the Cambridge City Council in Massachussetts had announced a three-month moratorium on recombinant DNA research within its boundaries while its own committee investigated the issues and assessed the adequacy of the NIH controls. The Cambridge Experimentation Review Board (CERB), a citizens' committee created by the Cambridge City Council in 1976, was charged with reviewing the issues concerning recombinant DNA technology.

Public concern over accountability persuaded local and federal officials to incorporate public representatives into the regulatory process. This included the integration of public representatives into the NIH-RAC, local biohazards committees and IBCs, the title of which was changed from Institutional Bio*hazard* Committee (IBC) to Institutional Bio*safety* Committee (IBC). Major changes in the NIH Guidelines in 1978 affected the composition of IBCs, stating that at least two members (but not less that 20 per cent) of the membership of each IBC shall not be affiliated with the Institution and shall represent the interests of the community with respect to health and protection of the environment (*Guidelines*, 1978). The federal government also designed programmes, such as the NSFs Ethical Values in Science and Technology (EVIST), Science for Citizens and the Office of Technology Assessment (OTA), to assist the public in gaining access to technical information and evaluating the impacts of new technologies (Goldberg 1985).

The pressure of public opinion in the USA was successful in promoting a move away from the scientific domination of the regulation of recombinant DNA work. The reconstituted IBCs and the public participation programmes offered a limited but novel potential for public participation in making decisions about recombinant DNA work, for decentralizing regulatory authority and for the local determination of safety matters. However, the shift of regulatory power to the public domain was temporary and soon the scientists' pressure groups were to swing back the pendulum of change away from public participation and accountability yet again.

3.4   THE FALSE CONSENSUS

As the public debate on hazards threatened to provoke further restrictions on recombinant DNA work, many molecular biologists felt frustrated by the stultifying effects on their research of the detailed protocols demanded for each proposed recombinant DNA experiment. Furthermore, converting existing laboratories to raise them to the requisite GMAG or NIH-RAC physical containment standard was very costly and resulted in the delay, if not total abandonment, of some lines of research. Many scientists became disenchanted with what they perceived as essentially unnecessary rules of conduct and constraints on experimental design which they maintained bore no relation to the actual hazards of their work. Scientists in the USA, in particular, imagined that their work would be superseded by research pursued under less strict controls in other countries, and those responsible for national science policy feared that the USA would lose its lead in the field.

In 1977 the Committee on Genetic Experimentation (COGENE) was created by the Sixteenth General Assembly of the International Council of Scientific Unions, an international non-governmental scientific organization of eighteen autonomous National Scientific Unions. COGENE was a scientists' pressure group formed to promote the value of genetic manipulation and the need to reduce the level of regulation of scientific activities in this field through its contacts throughout the scientific community and in government agencies.

It was during this phase of the recombinant DNA debate that certain risk-assessment procedures were addressed. This was undertaken at three important scientific meetings: the Enteric Bacteria Meeting at the NIH in Bethesda, Maryland in the USA in August 1976; a workshop entitled

'Risk Assessment of Recombinant DNA Experimentation with *Escherichia coli* K12' held in Falmouth, Massachussetts in the USA in June 1977; and a meeting entitled 'US-EMBO' (European Molecular Biology Organisation) Workshop to Assess Risks for Recombinant DNA Experiments Involving the Genomes of Animal, Plant and Insect Viruses' held in Ascot in England in January 1978.

As a result of these three meetings the community of molecular biologists reached a new consensus that recombinant organisms are safe to handle and that there are no unique hazards associated with recombinant DNA research. The results of these three meetings were subsequently cited repeatedly in scientists' testimony, in policy documents, and in statements of officials in the Department of Health, Education and Welfare and the NIH in the USA to justify the claim that there was little cause for concern. In 1979 COGENE co-sponsored with the British Royal Society a conference on 'Recombinant DNA and Genetic Experimentation'. In line with the new consensus, the conclusion reached at this conference was that research involving genetic recombination was safe.

Susan Wright, an historian of technology at the University of Michigan, argues that certain social characteristics of the three scientific meetings cited above acted as 'social filters' which reduced the complexity of the recombinant DNA debate. She considers that the main factors instrumental in this process of 'social filtration' were the structure of the decision-making process including the sponsorship and organization of the three meetings; the range of scientific and political participation in them; the informal processes affecting the scope of the proceedings and the reporting of the results; and the restrictions placed on the agenda for discussion (S. Wright 1986).

The restriction of political representation at these meetings determined that recombinant DNA hazards were assessed and analysed in a limited context which was characterized by informal interests with the objective of abating the recombinant DNA controversy. Wright maintains that concerns about the potential hazards of conducting recombinant DNA work were dropped from the agenda, whereas claims that certain types of risk were minimal were included. In particular, although the perceived risks associated with recombinant DNA work included ethical, moral and technical risks, the discussion was restricted to certain technical aspects.

The most important restriction placed on the discussion of risk was the assumption that, although the NIH Guidelines permitted the use of other organisms as hosts for recombinant DNA, all recombinant DNA research would be conducted with *E. coli* K12.

A second major restriction was that risk-assessment would be limited to hazards to communities outside the laboratories through secondary spread. The persistent focus on the question of the conversion of *E. coli* K12 into an epidemic pathogen allowed other considerations, such as hazards to workers through primary exposure inside the laboratory, to be peripheralized in the debate.

A third restriction was the assumption that recombinant DNA activities would be conducted only in advanced industrial countries with adequate public health and sewage treatment facilities. The implication was that epidemics in such environments could not occur under any circumstances.

The strength of the consensus amongst scientists, that the risks of recombinant DNA research were at worst very much smaller than they themselves had previously believed and perhaps non-existent, increased following the demonstration that hybrid DNA molecules produced by *in vivo* genetic recombination of DNA from different organisms occurred in nature (Chang and Cohen 1977). Advocates of recombinant DNA research used this information to argue that the production of hybrid DNA is not an unprecedented tampering with the balance of nature.

This demonstration of the existence of *in vivo* genetic recombination of DNA from different organisms altered the debate over the hazards of recombinant organisms from a qualitative argument—in the absence of recombinant DNA technology they would never occur—into a quantitative argument of how often they would be produced.

On the one hand, it was argued that although hybrid DNA molecules are produced in nature, the number of potential hybrid DNA molecules that recombinant DNA technology could produce is so large that it is unlikely that all of them have occurred naturally. On the other hand, it was argued that although the number of potential hybrid DNA molecules is very large, the age of the planet is such that an astronomical number of hybrid DNA molecules will have been produced over the course of a billion and a half years of evolution and therefore it is unlikely that recombinant DNA technology will engineer hybrid DNA molecules that have not already been produced naturally.

The latter viewpoint suggested to advocates of recombinant DNA research that, acting under the existing Guidelines, the probability of inadvertently creating a novel, viable, biological, hybrid 'monster' capable of self-replication through recombinant DNA technology, was very low since 'most recombinants have already been tried in the evolutionary arena and have been found wanting' (Freifelder 1978). It was argued that hybrid organisms, even when they survive, are usually incapable of reproducing and that the random probability of deleterious exchanges must be vanishingly low.

That an event is a naturally occurring phenomenon is, however, insufficient grounds *per se* for assuming that it is safe or that it is desirable to engineer its replication. That we have survived against a background of natural radioactive decay does not mean that nuclear power stations are not hazardous to the surrounding ecosystem. Analogously, that life has evolved in the milieu of *in vivo* genetic recombination is insufficient grounds for having confidence that recombinant DNA technology will not create genetic recombinations with catastrophic implications, *In vivo* genetic recombination involves many systems of control of expression that might be lacking from *in vitro* genetic recombination. Moreover, recombinant DNA techniques are used to create mutiple copies of organisms with the same genetic modification whereas chance genetic recombination in nature only produces one such copy at a time.

The new consensus that recombinant DNA technology was less dangerous than had been supposed had an immediate effect on the perception of risk associated with recombinant DNA work. For example, the arguments which had been used to determine the new consensus were used by prominent scientists in the USA to lobby several concerned senators, including Senator Edward Kennedy. As a result, in September 1977 Senator Adlai Stevenson called on the Senate to put off legislation in order not to act in haste, and Senator Kennedy withdrew support from his own bill, accepting the view that the hazards of recombinant DNA work were questionable rather than imminent (Gershon 1983). Despite the weakness of the scientific evidence and the flawed logic of the arguments upon which it is based, the new consensus has been used successfully to defuse opposition to recombinant DNA work, dismantle strategies to increase regulatory constraints and safeguards and, as described in section 3.5 below, to obtain significant relaxations of the guidelines governing its use.

3.5   GUIDELINES RELAXED

The new consensus regarding the hazards of recombinant DNA technology formed the basis of the substantial relaxation in the revised NIH Guidelines issued in December 1978, drafted by a NIH-RAC which had two laymen among its sixteen members.

The physical and biological containment levels were retained. Each classification of experiments, however, was downgraded at least one level in physical and/or biological containment. In addition, five categories of experiments, amounting to one-third of the experiments covered under the 1976 Guidelines, were exempted entirely from the Guidelines. The six classifications of prohibited experiments were maintained but with the NIH Director retaining the authority to grant case-by-case exemptions to the prohibitions, while experiments using viruses which had previously been prohibited were now assigned to appropriate containment levels. The Guidelines were further relaxed in 1980 by revisions which downgraded additional categories of experiments and exempted others completely from regulatory control, and again in May 1981, when experiments using the bacteria *E. coli* and *Bacillus subtilis* and yeasts were no longer required to be registered.

In September 1981 the NIH-RAC scientists recommended to the NIH that the Guidelines be converted from the existing compulsory set of requirements, with which all federally-funded laboratories had to comply, to a voluntary code of practice. However, public opinion was demanding the introduction of yet stricter federal regulations which the NIH-RAC wished to avoid. In 1982 a compromise was reached whereby the Guidelines would remain compulsory but the containment categories would be based on 'pathogenicity risk- assessment' rather than the then existing 'evolutionary' scheme of hazards (Gershon 1983). The NIH Guidelines of 1976 in the USA and the GMAG Advisory Notes of 1978 in the UK required increasing stringency of containment the closer the DNA under manipulation resembled human DNA (see Table 3.1.1). This 'evolutionary' scheme of hazards implied, according to Sydney Brenner, a leading molecular biologist working at Cambridge University in England, that 'lion DNA is more dangerous that pussycat DNA'. Brenner proposed a 'pathogenicity risk- assessment' scheme which required that in order to constitute a hazard, a piece of DNA has to meet three requirements:

(1) Access—the probability that escaped organisms containing the foreign recombinant DNA will enter the human body and reach susceptible cells.
(2) Expression—the probability that the foreign DNA incorporated into the genome of an organism will express its 'normal function', such as secretion of a toxin, that the organism formerly did not have.
(3) Damage—the probability that a foreign DNA sequence will cause physiological damage in the body to which it gains access once it is expressed (GMAG 1978).

A major characteristic of the regulatory relaxation in the USA was the increasing delegation of responsibilities by the NIH to the IBCs (Bereano 1984). The newly formed IBCs were reconstituted in 1978 to provide a quasi-independent review of recombinant DNA work done at an institution. However, from 1980, when the Reagan administration began to appoint representatives to the NIH-RAC, the participation of public and scientific members who played an active and crucial role in formulating critical debates, substantially declined. By 1980, IBCs were largely composed of DNA researchers and their associates.

As a result of relaxation of the Guidelines, 97 per cent of recombinant DNA work in the USA is subject only to IBC approval without review by the Office of Recombinant DNA Activity (ORDA) of the NIH, less than 15 per cent of all recombinant DNA work requires IBC approval before research has commenced, and no federal regulation whatsoever is required for the overwhelming majority of recombinant DNA experiments (Bereano 1984).

At most institutions, the IBC has essentially no budget with which to operate and thus most IBCs have inadequate resources to support the training of laboratory personnel and conduct health surveillance and laboratory monitoring. One IBC has little opportunity to benefit from the experience of another and the Office of Recombinant DNA Activities (ORDA) of the NIH is inadequately staffed and has insufficient resources to provide support to the IBCs. Critics of the revised Guidelines contend that rather than increasing accountability and protecting public health, the revisions in fact protect those engaged in recombinant DNA work from public enquiry and regulation (Bereano 1984).

An ORDA survey was completed in the USA early 1983 in which chairpersons of 250 IBCs were sent questionnaires, of which forty-five responded (an 18 per cent return). Seven of the forty-five IBCs had stopped meeting at all whilst others meet only infrequently because almost all of the recombinant DNA experiments performed at their institutions were exempt from the Guidelines. Business that does arise is often dealt with only by telephone (NIH, ORDA 1983).

Pressure from the scientific community together with increased awareness of the commercial promise of the application of recombinant DNA techniques persuaded the UK government that the risks involved in genetic manipulation had been substantially overstated. In January 1980 Brenner's 'pathogenicity risk-assessment' scheme was officially adopted by the GMAG. The outcome of this change was that the majority of experiments were downgraded to physical containment Category I. Initially, the GMAG committee considered each recombinant DNA experiment on an individual basis. Following the regulatory relaxation, most experiments in Category I were passed by the local biological safety committee, required by the HSW Act. No prior notification to the GMAG was required; it was sufficient merely for the local biological safety committee to notify the HSE annually of new research retrospectively. At containment level Category II, advance notice of recombinant DNA work was to be supplied to the HSE but no advice need be sought. Only the two highest levels of physical containment, Categories III and IV, were to be inspected by the HSE.

The constituency and functions of the GMAG came under review in 1983. Initially, the committee of the GMAG met monthly and collected evidence from up to seven expert subcommittees, but in 1982 it met only three times and had only one subcommittee. British trade-union representatives on the GMAG clearly failed to consolidate their role in the formation of national science policy as it related to genetic manipulations and to establish their participation in effective decision-making on other science policy-making bodies, such as the Medical Research Council (MRC). In 1984 the GMAG committee with its representation of scientists, non-scientific employees, trade-unionists, employers and lay-persons was disbanded in favour of the Advisory Committee for Genetic Manipulation (ACGM), which is largely composed of scientists. Unlike the GMAG, which was responsible to the DES, the ACGM is directly responsible to the HSE and is thus not an independent body (see Denselow 1982).

The relaxation of guidelines pertaining to recombinant DNA research and the reconstitution of the bodies which oversee such research in the USA and in the UK has resulted in a devolution of power to the laboratory level, as a consequence of which many scientists have stopped considering whether their research poses any special risk (Yoxen 1983).

3.6   RISK-EVALUATION

It is argued that the regulatory relaxation of recombinant DNA work prompted by the new consensus arrived at in the late 1970s has been justified by the experience of the mid-1980s. A great deal of virtually unregulated recombinant DNA work has been conducted in laboratories all over the world, including in countries without modern sewage systems, and using bacterial hosts other than *E.coli* K12, and to date there have been no reported cases of laboratory or industrial accidents involving recombinant organisms. However, there has been little organized risk-assessment on the hazards of recombinant DNA work.

At the present time, many organisms, including microbes other than *E. coli*, as well as higher organisms, are being used as hosts for recombinant DNA. Some of these organisms are not debilitated and can survive outside of a laboratory environment where they have the potential to reproduce, migrate and possibly mutate. It is not possible to recapture harmful living organisms which are accidentally released into the environment.

Practically speaking, the use of P1 physical containment is tantamount to no containment at all because P1 containment only involves good laboratory practices which should be standard even for non-hazardous research. Paul Berg revealed that in his own P3 facility designed for virus research at Stanford University, almost everyone who entered that laboratory acquired substantial antibody titres, a sign of infection, to SV40 (monkey virus) within one year (Waneck 1985). Physical containment minimizes risks because it reduces exposure, but it is not a substitute for biological containment.

Biological containment involves the use of enfeebled hosts, whose ability to survive outside of controlled laboratory conditions is much reduced, and enfeebled vectors whose ability to move from one host to another is debilitated. *E. coli* K12 had been grown and studied inside the laboratory for many years and it was thought that it could not survive outside the laboratory and therefore was safe. The first genetically enfeebled derivatives of *E. coli* K12 that complied with the NIH-RAC biological containment category EK2 were certified as biologically safe by the NIH-RAC in 1977. However, *E. coli* K12, the debilitated bacterial host favoured by the community of scientists involved in recombinant DNA work, is not as enfeebled as is generally believed. When the survivability of *E. coli* Chi 1776, an even weaker strain of *E. coli* K12, was tested more than two years after *E. coli* K12 was approved as a safe EK2 host, it was found that *E. coli* Chi 1776 could survive in some human volunteers up to 500 times the rate mandated by the NIH guidelines (Levy *et al.* 1980).

The survival of these bacteria was facilitated by the presence of a standard antibiotic-resistance plasmid which is used as a vector in many recombinant DNA experiments. When humans and animals on antibiotic treatment are fed antibiotic-resistant strains of *E. coli*, including *E. coli* K12, their colons become colonized by these strains in large numbers for periods of time which are longer than just a few days (Cohen and Laux 1985). This is because antibiotic treatment depletes the normal intestinal microbial flora and so the antibiotic-resistant strain is able to colonize the gut without competing with the normal indigenous microbial flora. Risk-assessment experiments have demonstrated that *E. coli* K12 cells can survive in the human gut for six days, and when the antibiotic ampicillin is administered to humans, the bacteria have been observed to persist in the gut for up to sixty-nine days (Levy *et al.* 1980). The results of these experiments indicate that people on antibiotic therapy should not participate in recombinant DNA work until some time after treatment is complete, when their normal intestinal microbial flora has re-established itself.

The principal criterion used for designating vectors biologically safe is that the vector should not be able to be transmitted from one bacterium to another during bacterial mating and should not be, or only inefficiently be, mobilised by other plasmids. The first enfeebled plasmid vectors to be certified as biologically safe were pMB9 and pBR322, which gained NIH-RAC approval in the late spring of 1977. The ACGM/HSE Advisory Note on disabled host/vector systems provides an accredited list of host-vector systems and states that where a derivative of any of the listed 'key' vectors has been developed, e.g., by deletion or by insertion of a sequence(s) from other listed vectors, it may also be regarded as disabled when used in an appropriate host (ACGM/HSE/Note 2). Each newly-constructed plasmid should be tested and found to be non-transmitting, since even vectors which are derived from pMB9 and pBR322 may not share the transmission properties of those plasmids.

The biological safety of vectors should not be assessed in isolation from the living systems in which they will operate because external factors can alter their transmission properties. An example of this involves pBR322 plasmids, which are now known to be efficiently mobilised if the host cell also contains certain other plasmids. The environment of the host cell may also affect the transmission properties of the vector. An example of this is the transfer of recombinant plasmids from *E. coli* to normal microbes in the intestines of living animals. Although the level of transfer is low in healthy individuals, it has been observed that in animals undergoing antibiotic therapy there is a significant level of transfer of plasmids to the intestinal flora (Cohen and Laux 1985). At the present time there is a range of vectors for different cells, some of which are constructs of viruses and even hybrids of plasmids and viruses. 'Shuttle vectors' are especially constructed so that they can replicate and express their genes in both bacterial cells and the cells of higher animals and plants. Such vectors increase the risk of accidental transfer of genes from bacteria to higher organisms.

If recombinant host-vector systems can survive for prolonged periods in the human intestine, then they will be excreted into sewage. Investigators found that enfeebled *E. coli* strains used for containment survived in raw sewage at roughly one-half the rate of indigenous bacteria (Sagik *et al.* 1981). These studies apparently had no bearing on the NIH Working Group on Revision of the Recombinant DNA Guidelines (1981), which concluded that the hazards of using these hosts and vectors were minimal.

The new scientific consensus asserts that physical and biological containment measures eliminate the risks attendant on recombinant DNA work. However, even if physical and biological containment measures were much improved, they cannot entirely remove risk because they can be breached through carelessness, laziness, poor judgement, sabotage, over-competitiveness, accidents and limited economic resources. Epidemics, individual health hazards and environmental disasters can arise as the unintentional consequences of accidents caused by human fallibility and carelessness, or sabotage or deliberate fraud. One risk inherent in very competitive modes of scientific endeavour is that certain scientists have occasionally contravened regulatory controls.

Some medical microbiologists asserted that nothing novel could arise as a result of the use of recombinant DNA techniques that was more hazardous than existing pathogens, and that the well-tried practices of medical microbiology would be entirely adequate to contain any potential hazards. However, the 1973 smallpox incident in the UK referred to above was a result of the failure of laboratory workers to adhere to prescribed safety procedures. Moreover, other incidents involving the escape of pathogenic microorganisms from medical microbiological laboratories have occurred in the UK as a result of such lapses. For example, in the high-containment facility at the Microbiological Research Establishment at Porton Down, a laboratory worker became infected with Marburg Disease, one of the pathogens in the category of greatest danger. In another laboratory a porter died of typhoid because he had adopted, unbeknown to his superiors, the practice of repacking the cans of infected material that he was putting into the sterilizing apparatus. At Birmingham University in 1978 a non-laboratory technician was infected with smallpox through a defective ventilation system, and in 1986 a cache of smallpox virus was discovered in a deep-freeze in a corridor in the London School of Hygiene and Tropical Medicine. Similarly in the USA, by October 1987, two laboratory employees working with HIV, the virus responsible for AIDS, were known to have become infected with HIV as a result of separate laboratory accidents (Palca 1987a).

Another type of risk which should be addressed in assessing recombinant DNA technology is that category of risk, reminiscent of the dystopian scenario depicted in Alduous Huxley's book, *Brave New World*, arising from its combination with social engineering as a tool for the social control of people, through the creation of subclasses based on genetic classification. Another social risk is that its application will lead to increased emphasis on technological solutions to health problems, thereby deflecting attention from other goals that are essential for social progress. Risk-assessment should also address the consequences of the nefarious use of recombinant DNA technology, for example by terrorist groups or unscrupulous covert agencies. The violation of evolutionary boundaries by the creation of hybrid DNA molecules, through *in vitro* genetic recombination of pieces of DNA from different species and the transfer of genes from one species to another through human interference, is considered by some to be unethical (Sinsheimer 1979).

Scientists tend to apply only technocratic notions of efficiency and to define risks only in terms of demonstrable and quantifiable evidence of actual harm. The whole recombinant DNA regulatory process is infused with a belief that the only risks that require investigation are the technical ones related to the possible escape of virulent bacteria from the laboratory, an attitude that fosters a regulatory preoccupation with physical and biological containment levels and their interchangeability (Bereano 1984). However, physical and biological containment measures merely reduce the probability that a given hazard will arise. It is fallacious to equate a low probability of a hazard occurring with the removal of risk. Although the perceived risk associated with recombinant DNA work includes ethical, social, ecological, political and technical risks, the only potential research hazards acknowledged by Brenner's risk-assessment scheme, adopted in the UK and the USA, are those that are able to be overcome by the ability of scientists to control and manipulate the variables and laboratory environments of their experiments. Furthermore, Brenner's scheme is restricted within its own terms of reference because it only addresses the probability of the occurrence of risk, and neglects the magnitude or severity of a hazard should it occur, and ignores both the distributional aspects of the hazard and who is to be accountable.

The NIH guidelines and proposals reviewed by the NIH-RAC are extremely technical. In all advanced industrial countries evaluation of the risks associated with recombinant DNA research is increasingly performed by scientists. The majority of the NIH-RAC has always been molecular biologists, and when public representatives have been unable to express health and safety concerns in quantifiable terms and support them with experimental evidence, they have not been taken seriously by the majority of the NIH-RAC (Goldberg 1985). Moreover, many scientists perceive the whole regulatory process as an infringement on an area in which their authority has traditionally not been questioned.

Lay people tend to define technical issues more broadly than experts and to be more cautious in their assessment of potential risks. Whilst risk may be thought of as a generally objective measure of harm, the notion of safety should be understood as being subjective. So that, whereas risk-assessment analysis requires expertise in the procedures of technology assessment and the analytical concepts of social risk analysis, safety is the level of acceptable risk. This means that although the IBCs in the USA, for example, may be competent to assess safety, their composition and resources are incommensurate with their ability to assess risk (Bereano 1984).

Knowledge is a source of power to which not all persons have equal access. A significant increase in the awareness of the general public is required in industrial societies in order that the bias of scientism can be challenged seriously and the implications of new technology assessed and regulated adequately. With assistance from experts, disinterested members of the public who represent the community with respect to health and protection of the environment can be effective participants in making decisions about the risks inherent in the development of science and technology. The effectiveness of public pressure groups is dependent upon their ability to find scientists who are willing to co-operate with them by explaining technical issues. Unfortunately, with a few notable exceptions, very few molecular biologists have been willing to risk compromising their professional reputations with their fellow scientists by providing the technical assistance needed to enable environmental or community groups to argue their points of view effectively.

3.7   TECHNOLOGICAL DETERMINISM

The predispositions of a given society can allow some interests to dominate over others, creating a 'false consensus' in which resistance is subsumed in the socio-economic structure and functioning of the society. The new scientific consensus on the hazards of recombinant DNA work which emerged from the three meetings of scientists discussed above is such a false consensus, and was facilitated by the institutional context of decision-making and the informal processes that limited the scope of the issues placed on the formal agenda. In arriving at the new scientific consensus, the complexity of risks in the recombinant DNA debate was simplified until what remained on the agenda was a risk-assessment debate about whether or not it was possible for *E. coli* K12 to overcome existing physical and biological containment measures and escape from the laboratory; whether or not it was possible for *E. coli* K12 to mutate into a pathogen capable of causing an epidemic; and whether or not it was likely that such an escaped pathogen could spread through the human population in advanced industrial countries.

The new consensus that recombinant DNA research would not produce pathogenic substances or organisms was not adequately substantiated by experimental evidence, nor were such experiments demanded. It would seem that the immediate concern of scientists at the centre of the recombinant DNA controversy after Asilomar was neither public safety nor scientific evidence, but that their freedom of investigation take precedence over the competing needs of the public and the health and safety of laboratory workers (Wright 1986).

The new consensus produced by the community of molecular biologists effectively peripheralized critics of the regulations. The environmentalist movement was accused of disregarding scientific evidence in pursuit of its opposition to all recombinant DNA work. Elliot Gershon, for example, wrote:

The radical opposition to recombinant DNA research should be judged harshly for the political agenda behind its ostensibly environmental concerns, and for its gross failure to evaluate correctly the actual dangers and benefits of recombinant DNA research. . . . The worst thing that could have happened in 1977 . . . was for legislation to be enacted [which] by its very existence would have been a triumph for the cataclysmic fears and the political ideology of the left (Gershon 1983).

Typically, in this way opponents of recombinant DNA technology were politically peripheralized by accusations against them that they were antithetical to science and technology and associated with the extreme left wing.

Following the new consensus, there has been a global trend in relaxing guidelines governing recombinant DNA research, and in the USA pending legislation aimed at regulating recombinant DNA work was removed from the political agenda. In the revised Guidelines in the USA there was a major reversal in the principles underlying recombinant DNA Guidelines: the burden of proof was transferred from scientists to show that recombinant DNA technology was safe to the general public to show that it was dangerous. Following the relaxation of the Guidelines there was a great expansion of the use of recombinant DNA technology in research resulting in a rapid expansion of knowledge in the field of molecular genetics and microgenetic engineering. Recognition of the commercial potential of recombinant DNA technology has encouraged increasingly close links between academic researchers in molecular biology and private companies.

Critics of the NIH-RAC argue that it has become a channel for those who promote the new technology rather than a forum for active debate among those with conflicting points of view (Goldberg 1985). Some of the most vocal participants on the NIH-RAC have major appointments in biotechnology firms or are actively promoting genetic engineering research. As Sheldon Krimsky, head of the International Network on the Social Impact of Biotechnology, has observed, 'the economic determinants of research and their influence. on the latitude of enquiry are both pervasive and subtle. Sometimes this influence manifests itself in the distortion of science. Other times it is expressed in the control of information. Most frequently it is felt by the kinds of questions that are pursued in the areas where science and social policy intersect' (Krimsky 1985).

The recombinant DNA debate has been subjected to a form of technological determinism in which the development of the technology itself is no longer in question. Jacques Ellul in *The Technological Society* has suggested that *la technique*—techniques and technical value systems—becomes an end in itself and that it is no longer within people's capacity to control technological developments (Ellul 1965). However, citizens of the city of Cambridge, Massachussetts, established their right to be involved in the decision-making process relating to the local development of recombinant DNA technology. In 1976 the Cambridge City Council created a citizens' committee, the Cambridge Experimentation Review Board (CERB), charged with reviewing the issues arising from the use of recombinant DNA technology. In their view:

Knowledge, whether for its own sake or for its potential benefits for humankind, cannot serve as a justification for introducing risks to the public unless an informed citizenry is willing to accept those risks. Decisions regarding the appropriate course between the risks and benefits of a potentially dangerous scientific enquiry must not be adjudicated within the inner circles of the scientific establishment (CERB 1977).

4 Jumping Genes

One of the major applications of microgenetic engineering is in research into the genetics of eucaryotic organisms. Prior to 1975, gene structure and expression in eucaryotic cells had not been much researched at the molecular level. Such research was impeded by regulatory guidelines and biological constraints, such as the complexity of DNA in higher-order life-forms, the inability to isolate specific fragments of DNA in large quantities and the shortage of suitable vectors. Regulatory guidelines restraining recombinant DNA research on eucaryotic organisms were relaxed as a result of the new consensus of the scientific community that the hazards of recombinant DNA technology had been overstated, and recombinant DNA techniques themselves became the tools used to overcome the biological constraints to eucaryotic genetic research. However, the resulting microgenetic analysis of eucaryotic organisms has increased rather than diminished cause for concern over the potential hazards arising from the use of recombinant DNA technology. In this chapter the evidence accumulating from this research is critically reviewed.

4.1 PROCARYOTES AND EUCARYOTES

When the cells of living organisms are examined under the light microscope, two basic types can be discerned: the procaryotic cell type and the eucaryotic cell type. Eucaryotic cells have a membrane-bound compartment within them, called the nucleus, in which the genome is sequestered on chromosomes. Procaryotic cells do not have a nucleus. Living organisms can be classified into two major groups on the basis of this difference in cell structure. The two groups are called the procaryotes and the eucaryotes. Higher plants and animals are eucaryotes and most bacteria and the blue-green algae are procaryotes.

The difference in size, complexity and the capacity for differentiation and evolution between procaryotic and eucaryotic cells is so great that until the fluctuation experiment of Delbruck and Luria (see below) bacteria were not credited by many scientists with having a hereditary mechanism at all. In procaryotic cells, the genes are arranged on a single molecule of double-stranded DNA, and procaryotic cells contain only a single copy of each gene. In the somatic cells of eucaryotic organisms, which contain vastly more genetic information than procaryotic cells, there are two copies of each gene. Procaryotic organisms usually reproduce by an asexual process of cell division, in which the daughter cells are normally identical to each other and the parent cell. Asexual procaryotic reproduction has more in common with the process of growth and maintenance in eucaryotes than it does with eucaryotic reproduction. Most eucaryotic organisms reproduce by sexual conjugation which results in the fusion of two specialized cells called sex cells which contain just half the usual amount of genetic information that is characteristic of the species. In contrast to asexual reproduction, sexual reproduction results in offspring which usually differ from each other and the parental type(s).

The chromosome theory of inheritance proposed that the Mendelian particles, the genes, are carried on chromosomes. However, it was not considered applicable to bacteria, which were widely held to be the last bastion of Lamarckism. In 1943 Max Delbruck and the naturalized American Salvador Luria, a medically trained microbiologist and refugee histologist from fascist Italy, conducted the so-called 'fluctuation experiment'. This now-famous experiment furnished statistical evidence of the spontaneous nature of bacterial mutation and was purported to be a direct confirmation of the theory of evolution by natural selection. The fluctuation experiment indicated that bacteria have a hereditary mechanism which is consistent with the theory of neo-Darwinism, and marked the birth of bacterial genetics. After the fluctuation experiment bacteria were widely adopted as the research material for the study of heredity. The relative genetic simplicity of bacteria and the rapidity with which they reproduce increases the speed with which genetic experiments can be carried out on them in comparison with higher organisms. From being considered so anomalous as to be external to the paradigm of classical genetics, the bacterial genome came to be regarded by many scientists as a model system for the study of eucaryotic organisms. In adopting the bacterial genome as a model for the study of heredity, the fundamental problem of biological reproduction was restated in terms of the maintenance and propagation of bacterial cells.

Most research on the molecular basis of genetics has been carried out on procaryotic organisms, particularly the bacterial species *E. coli* and on bacterial viruses. In the 1960s there was a growing interest in the study of gene regulation in higher animals and plants, which are eucaryotes. However, the extent to which this line of research was impeded by the inability to isolate single eucaryotic genes in large quantities prompted a number of scientists to leave molecular genetics to begin careers in neurobiology. What they had not foreseen was the rapid development of recombinant DNA technology, a new order of genetic manipulation which is able to overcome this limitation. Of particular use in eucaryotic research is a recombinant DNA technique called 'shotgunning', which enables fragments of DNA to be isolated and cloned (Watson *et al.* 1983).

In shotgunning, the DNA of the entire genome of an organism is cleaved into fragments. Each genome fragment is inserted into a plasmid vector which is then inserted into a bacterium. The plasmid replicates inside the bacterium and, in so doing, replicates the genome insert. The bacterium also repeatedly divides, quickly producing a colony of bacteria containing plasmids. Thus shotgunning is a technique for increasing the number of copies of a fragment of a genome.

The Guidelines for recombinant DNA research issued by the NIH-RAC in the USA in July 1976 signalled the end of the moratorium. It took until late 1976 to develop the first genetically enfeebled derivatives of *E. coli* K12 that would comply with the NIH-RAC biological containment category EK2, and these then had to be certified by the NIH-RAC as approved EK2 hosts. The first enfeebled plasmid vectors, pMB9 and pBR322, were certified as biologically safe by the NIH-RAC in the late spring of 1977. The NIH-RACs approval of these enfeebled hosts and vectors was interpreted by several European laboratories as a signal to proceed with recombinant DNA research, and structural analysis of specific eucaryotic DNA fragments using recombinant DNA techniques began in 1977. Although it was acknowledged that many details of the biochemistry of replication, transcription and translation in procaryotes were still unknown, it was assumed that the principles of molecular and biochemical genetics that had been formulated through procaryotic research, would be applicable to the genomes of higher organisms which had not been amenable to detailed analysis. However, the application of the techniques of microgenetic analysis to study the molecular genetics of higher plants and animals has revealed a profile of the eucaryotic genome which was unanticipated (M. Singer 1983; Watson *et al.* 1983).

4.2   INTRONS

In 1966, as a result of the molecular revolution in genetics, the manner in which DNA encodes information about protein molecules was deciphered. It was demonstrated that each amino acid is coded for by a sequence of three bases of the DNA molecule called a codon (see Chapter 1). Based on this information, the expectation was that there would be a correspondence between the size of proteins and genes such that a given gene would be comprised of approximately three times as many DNA bases as there were amino acids in the polypeptide chains of its protein product. However, it has been found that virtually all mammalian and vertebrate genes that have been analysed so far, and also the genes of eucaryotic micro-organisms such as yeast (though with much lower frequency) are larger than would be predicted if this assumption were correct. The reason for this is that the sections of genes that code for proteins are interrupted by intervening DNA sequences, called introns, that do not direct the production of any known protein (M. Singer 1983). Thus, in eucaryotic organisms genes are found to be split into pieces.

The number and size of introns, which are often considerably longer than the coding sequences, called exons, vary from gene to gene. For example, in the case of the human globin genes of haemoglobin molecules there are two introns, whereas in the gene for the structural protein alpha-collagen there are more than fifty introns.

The discovery of introns has prompted speculation on the evolutionary significance of split genes. There are diametrically opposed views on this, which include the hypotheses that they accelerate the rate of evolution or conversely that they retard it (Cherfas 1982).

Before a split gene is expressed as a protein, it is processed by the cell so that the introns are removed and the exons are joined together. The first stage of this process is that the entire split gene, including the introns, is transcribed into a long primary transcript of RNA. This RNA molecule then undergoes a series of steps in which the RNA introns are excised and the exons are spliced together, leaving a smaller mature messenger RNA (mRNA) molecule which consists of contiguous exons. The protein is then translated from this mature mRNA molecule.

The excision of introns from precursor mRNA molecules is a very precise process and the mechanism for its occurrence is not found in bacteria, whose genes are not split. In addition, it has been discovered that the regulatory sequences that control the expression of eucaryotic genes also differ from those in the bacterial genome and are in most cases non-functional when placed inside bacteria.

One of the issues raised in the recombinant DNA debate

in the 1970s was the potential of 'shotgun' cloning, in which random pieces of DNA from eucaryotes are cloned in bacteria, to create dangerous new pathogens if the eucaryotic DNA were expressed. The structural and regulatory differences which exist naturally between the genes of bacteria and the genes of eucaryotes makes the inadvertent expression of eucaryotic genes in bacteria unlikely. Indeed, this discovery is now used as a retrospective justification for the relaxation of regulatory constraints on recombinant DNA work. This natural difference in structure between procaryotic and eucaryotic genes was one of the initial technical hurdles in the application of recombinant DNA technology for the cloning of human protein molecules, such as insulin, in microbes. However, using microgenetic engineering, eucaryotic genes have been restructured to resemble and function like procaryotic genes. Introns have been removed, and these restructured genes have been joined to specially designed expression vectors constructed to facilitate the expression of eucaryotic genes in bacteria. In this way, gene transfer technology has surmounted the natural barrier to the expression of eucaryotic genes in procaryotic cells and, as a result, *E. coli.* has been made to synthesize eucaryotic proteins.

Those who argue against the possibility that the cloning of foreign genes in microbes could generate novel dangerous new pathogens, maintain that this is unlikely to occur because life has existed for such a long time that all possible genetic combinations will have already been tried in the evolutionary arena and found wanting. However, the ingenuity of the microgenetic engineer in violating the natural genetic barrier between procaryotes and eucaryotes has created a new order of genetic alchemy which does not exist in nature. In expressing a particular eucaryotic gene, a microbe is endowed with new metabolic capabilities which transform it into a microbe of a type which has probably never existed before and which may be harmful to the biosphere. The success of recombinant DNA technology in adjusting eucaryotic genes so that they are expressed in procaryotes has realized one of the potential hazards raised in the recombinant DNA debate of the 1970s. Without specific risk-assessment experiments, it is impossible to predict which eucaryotic genes have the potential to transform harmless bacteria into dangerous pathogens. However, few specific risk-assessment experiments to test such hazards have yet been conducted (Krimsky 1982).

4.3 ONCOGENES

Recombinant DNA techniques have also been applied in the study of cancer. In normal growth, development and body maintenance in higher animals, cell division is controlled so that the cells of one tissue multiply quickly enough to replace cells which are lost, but do not multiply so quickly that they invade other tissues or disrupt the way that an organ functions. A tumour is a clone of cells, derived from one progenitor cell, which has escaped the influence of factors which control the multiplication of normal cells.

The term oncogene (Gr. *onkos*: tumour) was coined in 1969 by Robert Heubner and George J. Todaro, of the Laboratory of Viral Carcinogenesis at the National Cancer Institute in the USA. Oncogenes are genes which are capable of transforming normal cells in culture to malignancy (Land *et al.* 1983). Two sets of oncogenes have been isolated. One set has been isolated from the genomes of tumour viruses, which are viruses associated with the onset of cancer, and the other set of oncogenes has been isolated from tumour cells. In addition to these, genes which closely resemble oncogenes have been isolated from the DNA of normal cells. Such genes, called proto-oncogenes, are believed to be a broad class of regulatory genes whose protein products control the activity of other genes and are crucial to the regulation of a cell's life-cycle. The proteins encoded by proto-oncogenes probably act at various key points in the control of cell division and play a role in the rapid multiplication and differentiation of cells during normal growth and development. Proto-oncogenes are found in the DNA of all eucaryotes.

The oncogene theory of cancer is that in proto-oncogenes we each carry the genetic predisposition for cancer in each of our cells. The transformation of cells into cancer cells is believed to be associated with the expression of a proto-oncogene at an inappropriate stage of development, or at an inappropriate level, or the expression of a mutated variant of a proto-oncogene, called an oncogene.

From the early 1970s it was suspected that certain viruses play a role in the initiation of human cancer. With the advent of new recombinant DNA techniques came a method of isolating the genomes of such viruses for molecular analysis. The NIH Guidelines of 1976 precluded the use of recombinant DNA procedures to study viruses associated with cancer. Following the significant relaxation of the Guidelines in December 1978, effective work on cancer-associated viruses commenced in 1979.

The theory of cancer-associated viruses in humans has steadily become stronger. Epstein–Barr virus has been found to be associated with Burkitt's lymphoma and with nasopharyngeal cancer; hepatitis-B virus DNA is found in hepatocellular carcinomas; human papilloma virus is likely to be found in all forms of human cervical cancer; and human T-cell lymphotropic viruses (HTLVs) are retroviruses that cause human leucaemias or lymphomas (Palca 1986).

Some viruses are believed to initiate cancer by producing a protein, called a promoter, which influences the rate of transcription of proto-oncogenes. Other viruses cause cancer through the possession in their genome of mutated versions of proto-oncogenes, called viral oncogenes. It is proposed that the small differences that exist between viral oncogenes and proto-oncogenes lead to significant changes in their protein products and that this creates specific cancer cells.

Retroviruses that possess a viral oncogene encoded in their genome are called highly oncogenic retroviruses because within weeks of infection they transform healthy cells in culture into cancer cells and induce tumours in host animals. One proposed mechanism for this cancerous transformation is that a DNA copy of the genome of the infecting retrovirus integrates into the DNA of the host cell where the viral oncogenes are translated and large amounts of their abnormal protein product is produced, which transforms the host cell into a cancerous cell.

Molecular biologists are cloning cancer-derived oncogenes and analysing them and their protein products in detail. *E. coli* K12 bacteria have been microgenetically engineered to contain and express oncogene DNA so that as much as 10 per cent of the total cell protein of the microgenetically engineered bacteria is oncogene protein. The study of oncogenes is an area of research activity which was only developed after the relaxation of the NIH Guidelines in 1980; consequently the recombinant DNA risk-assessment which was carried out in the 1970s did not address the risks of oncogene research (Bartels 1984).

The NIH Guidelines allow genetically engineered *E. coli* K12 bacteria with oncogenes inside them to be handled at the minimal physical safety containment level, P1, whether the work is conducted in a laboratory or on an industrial scale (Waneck 1985). It is not known how likely it is for an individual to become infected with *E. coli* bacteria which manufacture oncogene protein, nor is it known whether the colonization of the human gut by such *E. coli* constitutes a health hazard. It is necessary that researchers assess both theoretically and experimentally whether *E. coli* which have been microgenetically engineered to produce oncogene proteins constitute an occupational health hazard.

Another unknown risk arises from the possibility that plasmids used in the production of oncogene proteins would be transferred out of the genetically engineered *E. coli* K12 cells and into *E. coli* resident in the human gut. Risk-assessment experiments on the NIH-approved vector, bacterial plasmid pBR322, have not detected any such transmission (Levy *et al.* 1980). However, although they are derived from pBR322, the cloning plasmids which are used to achieve high levels of oncogene protein production are new constructs with a variety of novel properties. For example, efficient expression of one particular oncogene (*ras*) has been achieved in a hybrid plasmid constructed by combining the bacterial plasmid pBR322 with the tumour virus SV40. Quite unlike pBR322, this new plasmid can be propagated not only in bacteria but also in monkey cells growing in culture (McGrath *et al.* 1984). It is likely that the transmission properties in the gut for the newly constructed oncogene plasmids will be novel compared to those of pBR322. Therefore risk-assessment experiments to ascertain the transmission properties of each newly constructed plasmid should be conducted.

The cellular products of oncogenes are oncogene proteins, and it is these that are believed to be involved in carcinogenesis (Stacey and King 1984; Lautenberger *et al.* 1983). Work with animals other than humans has supplied evidence that contact with oncogenic DNA can lead to tumours. Such evidence is cause for alarm regarding the potential risk to workers handling naked oncogenic DNA. The expression of oncogene DNA in the cells of workers may transform those cells into the cancerous mode of growth, creating a potential health hazard to workers working with oncogenes. Rather than rely on casual discoveries of independent researchers or await the occurrence of cancer in laboratory workers, it is essential that the dangers to those workers who come into contact with oncogene DNA, with bacteria containing oncogene DNA and with oncogene proteins in research laboratories and in industrial production facilities be assessed on an experimental basis.

4.4 TRANSPOSABLE GENETIC ELEMENTS

In contrast with bacterial genes, which are contiguous on the bacterial DNA molecule and can be 'read' directly by bacterial enzymes, eucaryoctic genes are found to be separated by vast stretches of spacer DNA of unknown function. Indeed, approximately 90 per cent of the DNA in eucaryotic genomes does not appear to code for proteins and it is debatable whether it is involved in the translation of DNA into proteins. Amongst these DNA sequences of unknown function are certain DNA sequences which exist in multiple copies. For example, in the human genome there are between 300,000 and 1,000,000 copies of DNA elements of this type called the Alu family. It is possible that families of genetic elements provide raw material for the evolution of new genes with new functions.

The discovery that the majority of DNA in eucaryotic genomes is not involved in the expression of genetic information as structures in the phenotype has prompted a reconsideration of the definition of the gene. In the era of classical genetics the gene was conceived of as a functional unit. With the elucidation of the structure of DNA and the deciphering of the genetic code, the gene came to be considered as a structural entity—a sequence of bases on a DNA molecule. Genes ceased to be conceptualized as 'beads on a string' and came to be thought of as a linear sequence of nucleotide bases. The discovery that approximately 90 per cent of the DNA in eucaryotic cells does not appear to code for proteins has meant that once more a functional consideration must form part of the definition of the gene. Genes are now considered to be just those parts of the genome which encode the structure of, or play a role in the synthesis of, proteins, and they account for as little as 10 per cent of the DNA in eucaryotic cells.

Molecular genetics inherited the conceptual framework of a stable genome from classical genetics. It was not surprising, therefore, that the first reports of mobile genetic elements, or 'jumping genes' as they were called, deriving from McClintock's pre-molecular analysis of maize genetics, were mostly ignored until their existence was confirmed at the molecular level through molecular genetics studies on microbes in the 1960s which led to observations of moving pieces of DNA called transposons (Keller 1983). Whereas McClintock's concept of mobile genetic elements was functional, the molecular geneticists' concept of mobile genetic elements was structural. However, it was noted that transposons closely paralleled the properties of McClintock's controlling elements in maize (Fedoroff 1984). By the late 1970s transposons had been isolated from eucaryotic organisms, including *Drosophila* (fruit flies), yeasts and nematode worms and the term 'transposable genetic element' was introduced to describe all genetic elements that move from site to site within genomes.

Transposable genetic elements move around genomes by a process called duplicative transposition, which resembles the manner in which certain viruses replicate (Doolittle

1982). In duplicative transposition, a copy of a transposable genetic element located on a chromosome is duplicated and then deposited at a new location without loss of the original sequence.

The integration of a transposable genetic element at a new site may modify the expression of other genetic elements within the host cell, for example changing when and in which cells genes are active by turning the expression of genes on and off. Duplicative transposition can promote restructuring of the genome at various levels, from small changes involving a few DNA base pairs, to gross modifications such as inversions, omissions and duplications involving large segments of chromosomes (M.Singer 1983; Fedoroff 1984).

Sometimes transposable genetic elements travel as free DNA circles before integrating back into the chromosomal DNA at a new location. Others are transcribed into a free RNA intermediate molecule which, like a retrovirus, is copied through reverse transcription into a complementary DNA molecule which integrates into the chromosome. Sometimes during duplicative transposition a transposable genetic element is replicated many times, resulting in multiple copies of the genetic element. Certain genomes may consist almost entirely of transposable genetic elements or their inactive descendents (Rose and Doolittle 1983). Families of genetic elements, for example, are believed to have originated through the duplicative transposition of genetic elements. Families of genetic elements may also be expanded by reverse transcription of mRNA molecules to create copy DNA (cDNA) molecules which integrate into the genome. RNA introns are now believed to be a class of transposable genetic element with a role in communication between cells (Scott 1986a).

Barbara McClintock's interpretation of the mutagenic activity of the transposable genetic elements she observed in maize was that she was witnessing a disturbance of the orderly programmed response of the genome to challenge (McClintock 1978, 1984). One form of challenge is invasion by foreign agents to which animals respond by generating antibody diversity through rearrangement and mutation. When an animal is infected with a virus, bacterium or some other foreign body or chemical, the animal recognizes it as foreign and acts to remove it or destroy it by a process which involves producing proteins called antibodies. Each different foreign agent stimulates the production by the body of a different antibody. There are a vast array of microbial agents and an unlimited range of synthetic chemicals, most of which the animal has never encountered before, yet it has sufficient different antibodies to combat each, and each type of antibody has its own unique amino acid sequence. The 'paradox of antibody diversity' was how could an animal produce a unique antibody against a foreign agent that it, and possibly none of its ancestors, had ever been exposed to before?

One theory was that the enormous diversity of antibodies was inherited. Given that several million different antibodies have been characterized, each of which is coded for by a separate gene, this theory implied that a large fraction of the DNA in vertebrate cells would have to code for antibody molecules. The answer to this riddle only emerged in the mid-1970s through analysis using recombinant DNA techniques. It is now believed that only a few antibody genes are inherited and that the enormous diversity of antibodies is the product of an elaborate process of rearrangement, mutation and selection of these few genes (Tonegawa *et al.* 1978; see also Watson *et al.* 1983).

Molecular genetics studies have provided incontrovertible evidence that the concept of a static and stable genome was inaccurate. It would appear that the mobility and mutation of certain elements within the genome of vertebrate animals is normal and essential for the success of their immunological defences against invasion by foreign agents, and transposable genetic elements are now believed to be a major feature of all DNA (Shapiro 1983; Watson *et al.* 1983).

4.5 THE GENOME AS AN ECOSYSTEM

In addition to genes and an array of possibly non-functional genetic elements, the evidence of whose mobility was initially rather cautiously conceded by the scientific community, the genomes of eucaryotic cells contain other mobile genetic elements. These other genetic elements, which are integrated into eucaryotic genomes, are plasmid genes, dormant viruses and subviral infectious agents (Scott 1987, 1986a; Rose and Doolittle 1983; Watson *et al.* 1983).

Plasmids are circular DNA molecules found inside certain bacteria where for most of the time they are freely replicating circles of DNA. They are intimately linked to the survival and evolution of bacteria. They act as 'prime movers' in bacterial conjugation. From plasmids, some bacteria derive genetic plasticity and sexuality. Together with mutation, plasmids play a fundamental role in the processes of evolution in bacteria by enabling them to acquire genes from other bacterial strains. For example, multiple drug-resistance in bacteria results from the carriage of drug-resistance genes from one strain of bacterium to another on plasmids (see Day 1982). The Ti plasmid of the bacterium *A. tumefaciens* is able to integrate its genes into the genome of plant cells where they subvert the plant cell's metabolism and induce the formation of growths called gall tumours (see Watson *et al.* 1983). Moreover, it is speculated that *A. tumefaciens* may transfer other plasmids to plants and that, via plasmids, plants have access to the bacterial gene pool. It is suggested that through plasmid transfer, bacteria may have played an important role in plant evolution by transferring specific DNA elements to particular eucaryotic cells (Lichtenstein 1987).

Viruses are composed of a nucleic acid molecule encased in an outer coat. Within the DNA of eucaryotic cells are dormant viruses. Dormant viruses are DNA molecules integrated within the eucaryotic genome which have the latent capacity to initiate the production of virulent viruses. Unless the dormant virus is activated, the cell carrying it usually remains healthy. Genomic disturbance can mobilize dormant viruses inciting them to mobility within and without the genomic ecosystem. Once activated, the viral DNA switches from its dominant lysogenic phase to the lytic phase of its life cycle in which it converts the host cell into a mini-factory for making multiple replicas of itself, each of which is wrapped in an outer coat to form an infectious virus. These viral replicas burst out of the cell spreading infection to other cells and causing disease. A disturbance which catalyses the activation of a harmful virus dormant in the genomes of many members of a breeding group could lead to widespread pathological effects or even the extinction of the breeding group. One theory proposed for the spread of the human immunodeficiency virus (HIV) in Africa is that it was triggered by the mass-vaccination programme against smallpox conducted by the World Health Organisation (WHO) in the 1970s. In at least 50 per cent of cases, infection with HIV causes a severe deficiency of the immune system called acquired immune deficiency syndrome (AIDS). After infection with HIV a person can remain healthy for years and exhibit no symptoms of AIDS because the virus can lie dormant. It is only when the dormant virus is activated that the symptoms of AIDS develop. It has been postulated that dormant HIV viruses in the genomes of Africans innoculated with living vaccinia virus as a live vaccine against smallpox were activated thus causing the AIDS epidemic (Wright 1987).

It is a misconception to differentiate sharply between viruses and plasmids. In a similar way to plasmids, viruses are agents of gene transfer between cells. Occasionally, as a result of imperfect excision of a dormant virus from the host genome, host DNA is excised together with viral DNA and is packaged into the newly formed viruses. In this way, viruses are able to 'transduce' genes to another cell. The non-viral DNA thus introduced may alter the properties of the new host. Transducing viruses which infect organisms from more than one species may violate the breeding barriers that normally isolate the genes of one species from all other species. Evidence from the molecular analysis of classes of molecules common to different species suggests that, as the agents of interspecies gene transfer, viruses have played a considerable role in the generation of genetic variation – the platform upon which evolution can occur (Scott 1986a). At least six genes have so far been identified as strong candidates for past participation in such events, based on otherwise inexplicable similarities between genes of distantly related species. For example, a plant globin gene has been identified in leguminous plants which may be derived from the vertebrate gene family that codes for the haemoglobin in blood (Scott 1986a).

The third category of exogenous mobile genetic element which can be integrated into eucaryotic genomes are subviral infectious agents. As their name suggests, subviral infectious agents are simpler than viruses in structure. They are nomadic nucleic acid molecules that can enter cells and multiply within them. They may or may not become enclosed within an outer coat as they journey from cell to cell. Examples of subviral infectious agents are satellite viruses, viroids, virusoids, virogenes and virinos (Scott 1986b).

Through interacting with the structure and function of cellular genetic material, plasmids, viruses, subviral infectious agents and cellular mobile genetic elements have the potential to influence the fate of cells, organisms and species. For example, it is possible that the restructuring of the genome promoted by the activation of a dormant mobile genetic element could initiate cancer by, for example, altering the structure or level of expression of a proto-oncogene (Leder *et al.* 1983). A relatively small alteration of the genome of a fertilized egg having an effect early in embryogenesis could result in significant morphological changes giving rise to congenital abnormality. Genetic disturbance of the germ cells results in heritable variation. An *en masse* disturbance of the germ-line of a population by the mobilization of a dormant virus common to many members of a species could generate the requisite variation for a saltatory evolutionary event (Scott 1986a). It is speculated that in some cases speciation, which is the divergence of a species into groups which are unable to breed with each other, is the result of the spread of a virus or subviral infectious agent through the genome of the germ-line of a breeding group. Infection with a virus or subviral infectious agent could cause speciation by disrupting the structure of the genome in such a way that a fertile union between members of what was previously the same species would be prevented (Rose and Doolittle 1983).

The boundary lines drawn up by the communities of

scientists in genetics and microbiology determined that plasmids and viruses were accepted as microbiological entities rather than as 'anomalies' to the classical model of the genome. The classical concept of the genome, as a stable and static blueprint of the organism, is inaccurate. By their very nature, eucaryotic genomes are dynamic and interactive. There is a growing body of evidence that suggests that eucaryotic genomes are niches for a spectrum of genetic elements—DNA and RNA sequences—of a variety of 'life-styles'.

Genetic elements which inhabit the ecosystem of eucaryotic genomes include genes, non-coding genetic elements, plasmids, viruses and subviral particles. A comparison of the structures and characteristics of viruses, subviral particles, RNA introns, messenger RNA molecules, transposable genetic elements and proto-oncogenes reveals similarities which betray common origins. Some transposable genetic elements are akin to retroviruses in nucleotide sequence and, like retroviruses, have a free RNA intermediate in the non-integrated phase of their life-cycle. The genomic RNA of retroviruses has characteristics of eucaryotic mRNA molecules, and the oncogenes of highly oncogenic retroviruses are closely related to proto-oncogenes of cells.

Infectious genetic elements, such as viruses and subviral particles, may evolve from genetic elements which escape from cells. Conversely, genetic elements within eucaryotic genomes may be infectious DNA sequences which, like dormant viruses, remain integrated in the host genome. The relationships between members of the spectrum of genetic elements which inhabit eucaryotic genomes may be historical. Alternatively, their relationship could be in the present, with eucaryotic genomes in a permanent state of flux with temporal interchange of genetic elements between cellular 'life-styles' and infectious 'life-styles' (Scott 1987).

4.6   RE-ASSESSING THE RISKS

The classical concept of the genome is being challenged. The genome is no longer considered to be static. As a result of the insertion of genetic elements from distant genomic regions or from the DNA of infecting plasmids, viruses and subviral particles, the genetic contents of eucaryotic cells are in a constant state of flux—a caducean crucible of life.

Plasmids and viruses, the vectors exploited in recombinant DNA technology, belong to a spectrum of genetic elements which inhabit the ecosystem of eucaryotic genomes. The genomic ecosystem mediates developmental processes. Its disruption can be harmful to the cell, the organism and the species. In exploiting the natural integrative properties of viruses and plasmids as vectors in recombinant DNA work, there is a risk that the genomes of the target organism or of other organisms will be disrupted with deleterious effects.

To a certain degree, genomic ecosystems are buffered from some of the disruption caused by harmful agents. For example, there exist DNA repair mechanisms which are able, to a limited extent, to restore DNA sequences that have been mutated. Restriction enzymes are part of bacterial cells' defence against invasion by certain viruses. Similarly, once incorporated within bacterial cells, certain plasmids and viruses protect their new habitat, of which they are now an integral part, from invasion by further subcellular mobile genetic elements. A further example is the fertilized egg, in which the integrity of the union of one sperm cell with one egg cell is protected from invasion by the genetic material of further sperm cells. However, although genomic ecosystems are robust, like all biological systems, beyond certain parameters they fail to maintain homeostasis and dramatic changes occur.

The very accommodation of vectors within eucaryotic cells is potentially dangerous. The integration of vector DNA into the genome requires that the genome be cleaved and then rejoined. In the process of vector integration, genetic mutation of the genome can occur. Conditions which stress the genome through mutation and chromosome breakage activate transposable genetic elements which can then initiate inversions, omissions, exchanges and duplications of large segments of chromosomes (M. Singer 1983; Fedoroff 1984). Genomic disturbance which affects the expression of a proto-oncogene so that it is expressed at the wrong stage of development, or in the wrong cell type, or at an inappropriate level, is believed to be associated with the instigation of cancer. Furthermore, the integration of the vector may activate dormant viruses and subviral infectious agents within the target genome, with unpredictable consequences.

Even if the initial infection and integration of the vector does not precipitate a disruption to the eucaryotic genome, its presence within the genome may be a latent threat to the well-being of the cell, the organism, the species and the biosphere.

First, there is the risk that a factor which is intended for, and appropriate to, the activation of the expression of genes of the vector, may result in the activation of the expression of genetic elements within the host cell genome, or *vice versa*, resulting in unanticipated events. For example, the promoter which controls the expression of genes in the vector may also affect the expression of genetic elements within the host cell with undesirable consequences (e.g. see Hayward *et al.* 1981; Cheah *et al.* 1986). Equally, genes in the host genome and genes in the vector may share an activating stimulus, such as challenge to the immune system, as is postulated in the case of the activation of HIV and the onset of AIDS (e.g. see Palca 1986; Marx 1986). Once activated, a vector could wreak deleterious changes in the functioning of the cell. It could also infect other cells of the organism, and the cells of other organisms of the same species or indeed of other species.

The use of microgenetic engineering techniques to analyse the genetic contents of eucaryotic cells has revealed an unexpected profile. Genes are found to account for as little as 10 per cent of the DNA in eucaryotic cells, and there is a level of instability and interaction which was wholly unanticipated.

The vectors exploited in recombinant DNA work for the purpose of introducing foreign genes into cells are members of the spectrum of genetic elements that inhabit the ecosystem of eucaryotic genomes. Vectors are designed to infect cells and to alter their metabolism. In using vectors in recombinant DNA technology, the microgenetic engineer alters the dynamics of the cell's ecosystem.

There is a grave risk that recombinant DNA vectors will have unanticipated effects in the designated target cells. There is also the danger that, by virtue of their mobile nature, they will escape from the designated target cells and infect other cells where they will interact with other genetic elements.

The dynamic nature of eucaryotic genomes was not widely appreciated at the time of the relaxation of the recombinant DNA Guidelines. The Guidelines have not been adequately adjusted to take account of this dynamism nor have risk-assessment studies been undertaken which address the dynamic nature of cells as a potential hazard in microgenetic engineering. Each transposable genetic element, integrated subviral particle, dormant virus, viral vector or plasmid vector inside a cell is a potentially volatile agent for genomic rearrangements in response to stimuli from within and without the cell. The exploitation in recombinant DNA technology of viruses and plasmids as vectors and of recombinant viruses as live vaccines increases the probability of initiating major alterations in the genomes of cells, organisms and species throughout the biosphere.

Part II
The Bio-Industrial Complex

5 Utopian Science

The perception of scientists as unfettered enquirers, impartially testing hypotheses and responsible only to their consciences and to the scrutiny of peers, is utopian. In reality, the researcher's choice of project and the manner in which scientific investigations are conducted is constrained by the availability of scarce economic resources, the quality of team leadership and the prevailing political structure.

In the advanced industrial countries a close association between science, government and industry has developed. Without government involvement, market economies tend to under-invest in invention and basic research and development because such activities are unprofitable or technically risky or because the end-product can only be appropriated to a limited extent (Arrow 1970).

Science policy, the policy of government funding of research and development, has become highly politicalized since the expansion of government-funded scientific research and development in the years following the Second World War. Demands for government resources are made by different laboratories and research groups and from competing fields of science. Leaders in each field have increasingly acted to protect their positions and the resources available to them by lobbying members of government in order to obtain sectional representation in the governmental process. This intensification of group bargaining activities by the scientific community acting in the political arena has revealed the true compromise of the ideals of science (Haberer 1972).

Much recent commercial interest in genetic engineering has emanated from government incentives and national and international schemes and initiatives, for example, by introducing appropriate legislation and tax incentives. A plethora of government reports published in the early 1980s recommended that increased research funds be made available to biotechnology-related fields and that help and encouragement be given in the capitalization of new biotechnological enterprises and in the co-ordination of resources.

The public funds devoted in the UK to research and development important to the application of biotechnology are second only to the USA and Japan. In 1981 the British government merged the activities of the National Enterprise Board (NEB) and the National Research Development Corporation (NRDC) into the British Technology Group (BTG). The BTG funds academics for the early development of promising ideas; it funds the development of products or processes in industry; and assists in the startup of new companies. The BTG is probably the most substantial technology transfer organisation of its kind in the world. BTGs approach is to give priority to the development of inventions by British companies and assist them in exploring licensing opportunities at an early stage in the developmental process. In the pharmaceutical sector, BTG has concluded licenses with eighty per cent of the world's largest companies. It has substantial involvement in the fields of biotechnology and genetic engineering. In the early 1980s the BTG had over forty investments designed to stimulate bio-innovations, incorporating finance for small companies, strategic investment with industrial partners and assistance in technology transfer from university research centres to industry (see Table 5.0.1). Schemes such as the BTG are, at least in part, a recognition that serious obstacles to innovation arise in the late development and early production phases.

Many scientists have welcomed the increased association between science, government and industry because it has aided them in obtaining the order of financial support necessary for the development of 'big science'. Some economists have even suggested that state and industry should be fused into a form of corporatism which would subordinate the informal sector of the economy to monopolistic control in order to consolidate the factors affecting technical change, such as scientific, technological and managerial knowledge, and economies of scale, which are believed to result in economic growth (Rothwell and Zegveld 1985). It has, however, been argued that such technology policies will ultimately lead to an intervention-breeding system, as industrialists seek profits increasingly through political lobbying and government subsidies instead of through producing and selling (Jewkes 1972). Furthermore, there is increasing unease amongst the general public and many individual scientists over the close association of science, economics and politics (Salomon 1973).

Since the mid-1970s, commercialization of research in biology has resulted in a new order of university–industry liaisons (see *Biobusiness World Data Base* 1983; Lappe 1985). So-called technology policy encompasses policies and institutions which determine the emergence and application of technology. Government technology policy in recent years in all Western industrial countries has included communications, education, basic research, patent laws, tax policy, investment grants to firms, tariff policy, regulation of monopoly practices and industrial restructuring, support for small- and medium-sized firms, and funding of the technical and scientific infrastructure. In order to attract expertise from academe into business to generate high-technology innovations, 'innovation centres' and 'science parks' are being established with close links with development laboratories, patent agents, finance institutions and government departments.

In this chapter, in the context of the developments in genetic engineering, we explore the contradiction between the idealized self-perception of scientists as disinterested, apolitical and fraternal, with their aims and the means of achieving them in essential harmony, on the one hand, and the fundamentally materialist and political imperatives of the scientific community and its representative institutions, on the other.

Table 5.0.1: *The organization of the British Technology Group (BTG)*

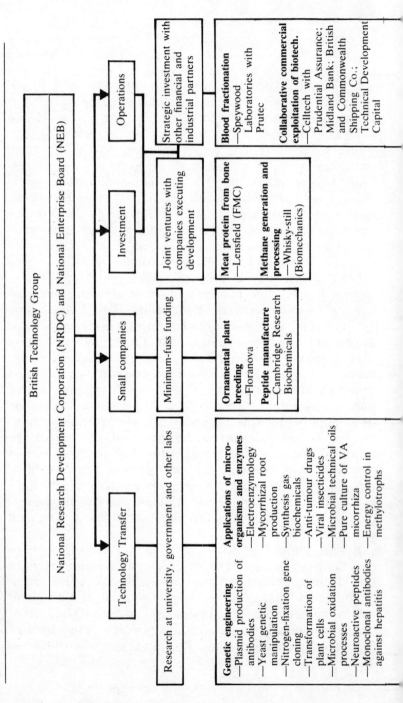

—Subunit rabies vaccine
—Starch-hydrolysing yeast
—Cell-transforming agents
—Cloning thermostable B-galactosidase
—Agrobacterium plasmid vector
—Improved strains of rhizobium
—Antibiotics from hybrid micro-organisms
—Immunodiagnostics for syphilis

Vaccines
—Caries vaccine
—Coccidiosis vaccine
—Veterinary respiratory disease vaccine
—Liver fluke disease vaccine
—Hepatitis vaccine

Antibiotics excluding genetic engineering.
—Antibiotics from skin flora
—Antibiotics from aquatic fungi
—Biosynthesis of penicillins and cephalosporins
—Biosynthetic studies on B-lactams
—Cephalosporin

New glucose-based protein
—Lyles with Rank Hovis McDougall
—Twyford Laboratories with Plant Resources Venture Fund

Source: D. Fishlock, The Business of Biotechnology, (London: Financial Times Business Information; 1982)

5.1   PATENTING IN GENETIC ENGINEERING

A particular economic incentive to substantial corporate involvement in molecular biology has been the granting of patents for recombinant organisms and engineered cell-lines. The purpose of a patent is to give its owner the legal right of action against anyone exploiting the patented research without the patentor's consent, and patentability can provide an impetus to innovation, research and development. In return for this monopoly on invention, patentors are legally bound to provide full technical details of their novel process or product, which are then published or otherwise made publicly available by each of the patenting authorities to which application for protection has been made.

There are three basic requirements for obtaining a patent; namely, the invention must be novel, it should not be obvious to an expert in the field, and it must have a potential industrial application. In addition to these requirements, the product or process must be patentable within the terms of the patent legislation.

Patentability depends on whether the invention is a product of nature or human-made. In the past, animal varieties and essentially biological processes for the pro-duction of plants and animals have been unpatentable, although special arrangements for protecting plant varieties exist in most countries. For example, the US 1930 Plant Patent Act and the 1970 Plant Variety Protection Act protect, respectively, plants which reproduce asexually and sexually reproducing plants. The UK 1983 Plant Varieties Act protects the reproductive material of plants, under the control of the Ministry of Agriculture, Fisheries and Food (MAFF) rather than the Patenting Office, and the 1968 International Union for the Protection of Plant Varieties is designed to achieve standardized protection of the reproductive material of plants (UPOV 1981).

The recent increase in the number of patents granted at the US Patent and Trademark Office (PTO) in 'mutation/ genetic engineering' (Fig. 5.1.1) is attributable to changes in the US PTO policy in view of the US Supreme Court

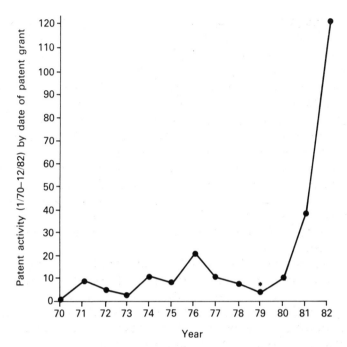

Figure 5.1.1 Patents granted in 'mutation/genetic engineering' by the US Patent and Trademark Office (PTO), 1970–82

Note: * In 1979 the US PTO issued 26 per cent fewer patents owing to financial constraints

Source: Includes all of the patents issued to US and non-US applicants from Subclass 172 of Class 435, and selected patents from the following Subclasses of Class 435: 5, 6, 8, 89–91, 253, 820; and selected patents from Subclasses 85 and 117 of Class 424 of the US Patent Classification System. Derived from *Patent Profiles: Biotechnology* (1982) (Washington DC: Office of Technology Assessment and Forecast (OTAF) for the US PTO).

decision in *Diamond* v. *Chakrabarty* (1980) (Saliwanchik 1982; Wheale and McNally 1986).

In this case, the patent application was filed in 1972 by Ananda Chakrabarty, a scientist employed by General Electric, for a genetically engineered form of the bacterium

Pseudomonas. The organism was developed using genetic engineering techniques (but not using recombinant DNA techniques) to degrade crude oil from oil spills in water. The bacteria would ingest the degradation products and then themselves form part of the food chain.

Patents for the process used in producing the bacteria and for the method of their dispersal in water were awarded by the Patents Examiner, but the application for a patent on the organism itself was rejected on the grounds that organisms like the one that was engineered occur in nature and are therefore unpatentable. However, the Court of Customs and Patent Appeals (CCPA) rejected this interpretation of the Patent Act. In 1979 the US government lodged an appeal to the Supreme Court against the CCPA's decision. The issue of debate in the Supreme Court in the Chakrabarty case was whether or not a human-made living organism is patentable within the meaning of the Patent Act.

Although Chakrabarty's micro-organism was not produced using recombinant DNA techniques, it was recognized at the time that a decision by the Supreme Court in favour of Chakrabarty would confirm the patentability of micro-organisms produced using recombinant DNA technology. During the course of the Court's deliberations, an *amicus brief* was filed by the People's Business Commission (now The Foundation on Economic Trends), a non-profit educational foundation designed to raise public awareness on economic and technical trends in the USA. The *amicus brief* filed by the People's Business Commission alerted the Supreme Court to the incentive for the increased commercial exploitation of recombinant DNA technology that the patentability of recombinant organisms would provide and, believing recombinant DNA technology to be potentially hazardous, urged a decision against Chakrabarty as the granting of the patent would not be in the public interest. To counter this argument, Genentech, one of the leading genetic engineering companies in the USA, also filed an *amicus brief* in which it cited the relaxation of the NIH Guidelines as evidence of scientific consensus on the safety of recombinant DNA technology (Rosenblatt 1982).

The Chakrabarty case was resolved by the decision of

the Supreme Court in 1980 that the US PTO can issue patents on live human-made micro-organisms, and indeed on any species of living thing, provided it can be shown to be a product of manufacture. The implication of this decision was that researchers who use recombinant DNA techniques to manufacture a new organism could protect their proprietary interests in that organism in the USA with a government patent. Following the Chakrabarty ruling, the US PTO directed its attention to the vast backlog of patent applications based on recombinant DNA technology which had been delayed pending the Chakrabarty decision.

In 1973 and 1974 the research teams of Stanley Cohen of Stanford University and Herbert Boyer of the University of California in San Francisco (UCSF) demonstrated the feasibility of using recombinant DNA techniques to engineer the replication of foreign, biologically functional genes introduced into a living organism—cells of *E. coli*—on a recombinant plasmid vector. In 1974 Stanford University's Technology Licensing Office filed a patent application for recombinant DNA techniques which included techniques for transforming cells with recombinant plasmids, using antibiotic-resistance genes on plasmids as genetic markers, *in vitro* genetic recombination techniques for creating recombinant plasmids and for the recombinant plasmids themselves. In 1978 the patent application was split into two patent applications—for a process patent and a product patent. Industrial analysts predicted that together these patents might be worth more than 1,000 million dollars (Walton and Hammer 1983). (See also 2.2.)

The first Cohen–Boyer patent, issued in 1980 after a delay of six years and much controversy, is called the 'process patent'. It describes the prototype recombinant DNA techniques which were used to splice the gene coding for human insulin into antibiotic-resistance plasmids of *E. coli*, and the insertion of recombinant plasmids into *E. coli* bacteria, creating novel life-forms, and the selection of recombinant bacterial colonies using the antibiotic-resistance genes on the plasmid vectors. The second patent, called the 'product patent', was issued after ten years of controversy, in August 1984. It covers products produced by bacterial plasmids in bacterial hosts.

Under the UK Patents Act of 1977, microbiological processes and products are patentable provided there is significant human intervention in the discovery of such micro-organisms, e.g., the isolation of a micro-organism from its environment. Thus, in principle, a patent can be obtained for: micro-organisms *per se* (including viruses); processes for producing micro-organisms; processes using micro-organisms; products obtained from microbiological processes; DNA/RNA molecular sub-cellular units (e.g. plasmids); and cell-lines in culture. There are many UK patents on record for recombinant DNA techniques and patents have been issued on recombinant microbes. Section 1.3.b of the UK Patents Act (1977) states that a patent shall not be granted for any variety of animal or plant or essentially biological process for the production of animals or plants not being a microbiological process or a product of such a process. A key word in the Act is *microbiological*, the interpretation of which will determine whether patents will be granted for applications for algae and the cell-lines of higher animals or plants. Similarly, until a ruling on a suitable patent application has been made, it remains to be seen whether the phrases *variety* and *essentially biological process* in Section 1.3.b of the Act will be interpreted to include novel animals and plants created using the new techniques of genetic engineering.

The decision to permit the filing of patent applications for genetically engineered animals and plants has already been made in the USA. In April 1987, the US PTO ruled that genetic engineers could patent the new species of plants and animals (other than humans) they create and by the end of that year more than fifteen patent applications for genetically engineered animals were on file. These included applications for patents on a chimaera (hybrid) between a sheep and a goat created through cell fusion, and a pig whose genetic structure has been altered to make it grow faster and produce leaner meat. In April 1988 the US PTO issued the first such patent to Harvard University. The patent covers any recombinant animal engineered to contain an activated oncogene.

There are a number of problems in obtaining a patent and enforcing patent protection, both peculiar to the field of genetic engineering and applicable to patenting in general, which are exemplified by the disputes surrounding the granting of the Cohen–Boyer patents. These include establishing the patentability of the subject-matter, proof of infringement and scope of protection, prior disclosure, proof of inventorship and adequate disclosure.

In most countries any discussion, even one day in advance of an application, invalidates the patent because of prior disclosure. There is, however, considerable variation between the patent regulations of different countries, including exceptions to the prior disclosure requirement, notably in the USA, Canada and Japan, where formal application is required within, respectively, one year, two years and six months of any public discussion.

An article discussing the research in the Cohen-Boyer process patent application was published in the British magazine *New Scientist* one year and one week prior to the filing of the patent in the USA. This prior disclosure wrecked its patentability prospects in Europe and seriously jeopardized the validity of the patent application in the USA (Sylvester and Klotz 1983). Stanford University was able to show that insufficient information had been revealed in the article to enable replication of the techniques by other scientists and the US PTO eventually issued the patent, but other national patent offices did not.

The issue of prior disclosure provides an example of the problems that arise when science is perceived of as a commodity. The implication here is that any scientific disclosure, even at a closed conference, may jeopardize patentability, and scientists may therefore decide to withold scientific papers from publication and refrain from discussion of their work. Indeed, Roman Saliwanchik, a patent lawyer in the USA, has advised scientists that if their inventions are likely to lead to effective patenting, then the patent application should be filed before the research is published (Saliwanchik 1982). With careful timing, the dilemma of the scientist, undecided between disclosure for acclaim or secrecy for profit, may be resolved in the USA, Canada and Japan, where the patent laws are less strict than in Europe, and where a scientist can be nominated for a Nobel Prize and may still apply for a patent.

Establishing a claim to a patent may be complicated if too many people are represented as having made a particular experiment possible. A dispute arose over the granting of the Cohen–Boyer process patent when it became clear that the development of recombinant DNA techniques had relied heavily on discoveries made at several institutions, for instance on work on restriction enzymes at Johns Hopkins University and by Paul Berg's group at Stanford University, and research on the functioning of ligases at the NIH Laboratories.

The clash between the scientific convention of co-authorship with the legal requirements necessary to substantiate claims of inventorship also delayed the granting of the first Cohen–Boyer patent. A key article published in 1973 cited in the patent application, describing the construction and replication of a plasmid capable of transferring anti-biotic-resistance into *E. coli*, names two other authors in addition to Cohen and Boyer, namely Annie Chang of Stanford University and Robert Helling, Associate Professor of Botany at the University of Michigan (S. Cohen *et al.* 1973). According to the rules of the US PTO, co-authors are considered co-inventors of the processes described. However, the US PTO procedures allow for the filing of the reissue of a patent to change named inventors.

Helling, who worked with Boyer at the UCSF on a sabbatical year during 1972–3, was not named in the patent application. He disputed that his role was marginal and initially refused to sign a disclaimer, required by the PTO prior to approval of the patent application, agreeing that he was not an inventor of the processes described. John Morrow of Johns Hopkins University, first-named author of a 1974 *PNAS* paper (Morrow *et al.* 1974) cited by Cohen and Boyer in the patent application, also refused to sign a disclaimer to the patent application. (See also 2.2.)

One of the basic requirements of patentability is that the invention or discovery must be novel. During the process of granting the second Cohen-Boyer patent, evidence came to light of a prior publication that brought into question the novelty and originality of Cohen and Boyer's work as defined by the Patent Act. Although Cohen and Boyer's research teams were the first to do the laboratory experiments, a PhD thesis by a Stanford University graduate student, Peter Lobban, published a year earlier in May 1972, set out the complete conceptual framework of the recombinant DNA process. Eventually, these particular disputes concerning the process patent were decided in favour of the Cohen–Boyer patent application and the US PTO issued the patent in 1980.

In the Cohen–Boyer process patent, Stanford University and the UCSF own US property rights until 1997 over the basic methodology and tools used throughout the genetic engineering industry (Sylvester and Klotz 1983). However, there is a problem in policing the use of the patented process. The process is used at the production level, and so infringement of the patent might be difficult to detect and prove. Another problem is that the Cohen–Boyer patented process would generally be practised only once in the course of the development of a new micro-organism. The value of the product patent is that, it is easier to police than the process patent. The patent claim would be directly infringed by any unlicensed US or foreign concern that makes, uses or sells a hybrid plasmid of any type in the USA.

The passage of the product patent through the US PTO was also controversial. In return for a limited monopoly on their invention, patentors are legally bound to provide full technical details of their novel process or product. This is known as the enablement requirement. When the manner of producing an invention can be expressed in words in sufficient detail to enable the claimed invention to be duplicated, then a written description is considered adequate disclosure to satisfy the enablement requirement. The problem of what constituted adequate disclosure arose initially in the patenting of antibiotics produced by micro-organisms isolated from the environment. The chemical structures of these antibiotics was sufficiently complex that no method of synthesis that would produce the desired product could be described. Even when the detailed taxonomy of the antibiotic-producing micro-organism could be supplied, it is doubtful that someone wishing to duplicate the invention could isolate the desired strain from the whole universe of micro-organisms. The only practical way to ensure that the public could reproduce the invention was to require that the inventor place the antibiotic-producing micro-organism in a depository from which it could be obtained by the public (Buting 1984). The Budapest Treaty of 1977, to which fifteen countries are at present signatories, requires that an organism for which patent protection is sought must be deposited in an officially designated international depository (*Budapest Treaty* 1981).

In the field of recombinant DNA technology, the deposit requirements of the Budapest Treaty may not apply in cases where the construction of the plasmid or other vector can be described adequately to comply with the requirement for sufficient disclosure. However, questions were raised by the US PTO over whether the plasmid pSC101 could be prepared in accordance with the working example described in the Cohen-Boyer product patent application. Neither was the plasmid deposited at an officially designated depository at the time of the patent application. In 1983 these objections of insufficient disclosure were dropped by the US PTO on the grounds that the plasmid was sufficiently disclosed because Cohen claimed that it had been available to any one who requested it, subject only to some safety guidelines and a request that it should not be passed on to others.

Legally defined constraints imposed by commercial considerations are proving a problem for academic scientists concerned to retain the integrity of the research process. There is a dilemma for scientists employed in the public sector who believe that the fruits of their labour should be freely available for the public good. Stanley Cohen was initially reluctant to apply for patents on recombinant DNA technology because he believed patenting would limit its application. In a similar vein, Alexander Fleming, who discovered penicillin in England in 1928, did not patent his finding because he believed the knowledge should be freely shared. The outcome of the penicillin story is that the method for mass-producing penicillin was patented by the Department of Agriculture in the USA (USDA), thus enabling the USA to exploit the innovations of basic research performed in Britain, and for which British researchers subsequently had to pay royalties to the USA (Sylvester and Klotz 1983).

Patents offer inventors a limited monopoly on their discoveries. However, to ensure that the public receives the full benefit of the new knowledge, in exchange for the granting of a limited monopoly to the patent holder, in the enablement requirement, patents make the invention or discovery available for use to anyone with reasonable skill in the field. The Budapest Treaty requires that if written description is unable to supply sufficient information for disclosure, an organism for which patent protection is sought must be deposited at the time of patent application, that is prior to patent protection. Thus, in the case of a micro-organism, unlike conventional inventions, the invention itself, rather than a description sufficient to enable a person of ordinary skill in that technology to make it, must actually be given to any competitor who desires it. The risk of industrial theft arising from the Budapest Treaty disclosure requirements is a deterrent to organizations and individuals from patenting their microbial inventions.

A trade secret contrasts with a patent. Many companies believe it is easier to keep their trade secrets secure than to defend patents against infringement. Where there are corporate-university relationships, companies could invoke the right of keeping information a trade secret rather than taking out a patent which would allow a person with ordinary skill in the field to duplicate the invention.

Most scientists agree that universities are justified in seeking to benefit from research pioneered by scientists employed by them, but an ethical difficulty arises in the granting of private licensing rights to the results of research carried out on public funds. As Sandra Blakeslee comments, 'Trust in universities could be eroded. Public funding of academic institutions has meant that citizens could turn to universities for unbiased, disinterested advice. When the advice is no longer disinterested, the public is cheated' (Blakeslee 1985).

The questionable morality of granting licensing rights to institutions for their research when it is carried out on public funds must be weighed against the possibility that private companies may obtain the materials or techniques which enable them to develop patentable processes or products, or indeed, having obtained such materials or techniques, keep them as trade secrets for the purpose of commercial exploitation.

5.2   SCIENCE AS A COMMODITY

The effect of treating science as a commodity is to impede the free flow of information and communication throughout the scientific community, thus undermining the informal tradition and ideals of the scientific enterprise (Gibbons and Wittrock 1985).

A dispute between the University of California in Los Angeles (UCLA) and the pharmaceutical manufacturer Hoffman La Roche over the microbial cloning of interferons concerned researcher's rights in cells and other materials exchanged under the protection of mutual trust and gentleman's agreements. Interferons are the body's natural anti-viral proteins. They are a class of proteins released by certain mammalian cells in response to various stimuli, including viral infection. In 1980 the gene for interferon was valued at several billion dollars on the strength that microbially cloned interferon would be an effective drug against cancer (Wade 1980a).

In 1977 David Golde and Philip Koeffler, two researchers at the UCLA, successfully sustained growth and cell division *in vitro* of cells extracted from the bone marrow of a patient dying of leucaemia. They named the cell-line KG-1. Golde and Koeffler supplied samples of KG-1 cells to Robert Gallo at the National Cancer Institute, who established that they produced interferon. Without informing Koeffler and Golde, Gallo made the cells available to Sidney Pestka, a researcher at the Roche Institute of Molecular Biology, an establishment wholly funded by the pharmaceutical company Hoffman La Roche.

Without the consent of Golde and Koeffler, Pestka, using genetic engineering techniques, succeeded in improving the interferon-producing capacity of KG-1 cells. Hoffman La Roche subsequently entered into an agreement with the

US genetic engineering company, Genentech, for the microbial production of interferon. In order to achieve this, Genentech made use of the interferon gene from the KG-1 cell-line. Hoffman La Roche and Genentech filed a joint patent application covering both the interferon made from the KG-1 gene and the process whereby the gene was spliced. The UCLA claimed that Hoffman La Roche and its contractor, Genentech, made unauthorized use of KG-1 cells for commercial purposes after they had distributed the material to colleagues in accordance with the tradition of scientific collaboration. After three years of legal wrangles an out-of-court settlement was reached at the end of 1982 for a large undisclosed cash sum paid by Hoffman La Roche to the UCLA. The UCLA–Hoffman La Roche dispute illustrates the strain that commercial potential places on the tradition of informal exchange of scientific materials and information.

Scientists are as competitive in their work as other groups in society. The rewards for achievement in scientific enterprise are accolades from academe and, increasingly, financial rewards from commerce and industry. In the competitive environment of international scientific endeavour, rewards often go to the scientist who can substantiate his or her claim to priority. It is this desire to be the first to make an important discovery or perform a pioneering experiment which sometimes leads a scientist to compromise standards of openness and scientific communication and increases the likelihood of accident and carelessness and there have been some instances when scientists have breached regulations on the type of experiment that may be conducted (Wade 1980b).

In 1980 Martin Cline, working at the UCLA, abandoned the usual ethical standards expected in medical treatment, when he attempted gene replacement therapy using recombinant DNA techniques to correct a defect in the bone marrow cells of two people suffering from the hereditary genetic disorder beta-thalassaemia. The attempt was unsuccessful, and in June 1981 the NIH withdrew their grant support from his research at UCLA as a punishment for his premature experimentation without informed consent (Wade 1981). Cline later admitted that he used 'poor judgement' in failing to adhere to the Guidelines and in misinforming the authorities of his intentions.

A major criticism of Cline was that the results of his experimental trials of gene therapy on mice gave no scientific grounds for confidence that such therapy would be efficacious on humans. In his haste to pioneer this innovative approach to therapy, Cline violated federal recombinant DNA Guidelines as well as federal regulations on human experimentation designed to protect the health and well-being of patients.

The alleged misuse for commercial gain of material specifically supplied for research purposes was at the centre of a dispute between a team of researchers led by Luc Montagnier at the Pasteur Institute in Paris and a team of researchers led by Robert Gallo, head of the laboratory of Tumor and Cell Biology at the US National Cancer Institute in Maryland. The controversy concerned the source of the virus upon which Gallo's research team based a commercially valuable test with which to screen blood for AIDS carriers.

In May 1983 Montagnier's research team published an article in the journal *Science* claiming to have discovered the virus responsible for acquired immune deficiency syndrome (AIDS) (Montagnier *et al.* 1983). The article contained a description of the virus together with electron microscope photographs of it. Montagnier's research team subsequently called the virus which they had discovered the lymphadenopathy associated virus (LAV). In July 1983 Montagnier's laboratory supplied samples of LAV to the research team of Gallo. Gallo claimed that he was unable to grow the samples of LAV which were sent to him and requested a further sample, which was supplied by the Pasteur Institute in September 1983, together with a contract which stated that the virus must be used for research purposes only and not for industrial and commercial purposes. Gallo claimed that he was also unable to grow this second sample of LAV.

Gallo's stance at this time was that the AIDS virus was not LAV but that the AIDS virus belonged to a group of viruses, the human T-cell lymphotropic viruses (HTLVs), upon which he had established his reputation. Initially,

Gallo maintained that the AIDS virus was HTLV-I or a variant of it. In April 1984 the Department of Health and Human Services (DHHS) in the USA announced that Gallo had discovered the virus responsible for AIDS and that it was an HTLV virus, not HTLV-I, but a different HTLV virus which Gallo called HTLV-III.

Following his 'discovery' of HTLV-III, Gallo was credited with having discovered the virus responsible for AIDS (see Marx 1984). Even before Gallo claimed to have discovered HTLV-III, the media had supported his claim that the AIDS virus was an HTLV virus (see Marx 1983). Little support was forthcoming for the work of Montagnier's team, who had actually isolated an AIDS virus and had demonstrated that it was not an HTLV virus. In May 1986 scientists on a subcommittee of the International Committee on Taxonomy of Viruses in the USA concluded that the AIDS virus did not belong to the HTLV family and decided that it should be called Human Immunodeficiency Virus, or HIV, yet despite this decision, Gallo still persisted in referring to the AIDS virus as HTLV-III.

The AIDS virus is constantly evolving new strains because it mutates as it replicates. The nucleotide sequence of the HTLV-III virus, which Gallo claims to have isolated, and LAV, the virus which Montagnier's team isolated and twice supplied to Gallo, are almost identical and more similar than any other two samples of AIDS virus which have ever been compared. This similarity suggests that Gallo's isolates of HTLV-III were somehow contaminated with the isolates of LAV that Montagnier supplied to him in 1983, a proposition which Gallo denies. There is, however, further evidence that Gallo's laboratory was growing samples of LAV that Gallo had claimed had failed to grow. In May 1984, when Gallo claimed to have discovered HTLV-III, he published three photographs of the virus (Gallo *et al.* 1984) which were later proven to be photographs of the French virus LAV. The Pasteur Institute's legal advisers believe the mistake over the photographs is evidence that Gallo had not isolated HTLV-III before December 1983 and provides strong circumstantial evidence that Gallo's HTLV-III virus is in fact LAV (Connor 1987b).

Tests have been developed at the Pasteur Institute in France and at the National Cancer Institute in the USA to screen for the presence of antibodies to the AIDS virus in blood and thereby identify carriers of the AIDS virus. The Pasteur Institute was the first to apply for patents on a test for antibodies to the AIDS virus when it filed an application for a British patent in September 1983. It also applied for a US patent in December 1983. On the same day that Gallo published his 'discovery' of HTLV-III in April 1984, the US government filed a patent on an AIDS test developed by Gallo and purported to be based on HTLV-III. The US PTO granted Gallo's patent in May 1985, whilst the Pasteur Institute's patent application in the USA was still pending. This enabled the US government to license their test for AIDS antibodies whilst denying the Pasteur Institute similar licensing agreements. The royalties on these test kits have amounted to millions of dollars (Connor 1987b).

The controversy surrounding the AIDS virus and its use in screening tests led to a dispute, based on three court cases, between the Pasteur Institute in France and the DHHS in the USA. The first court case concerned which institutional body had rights to a patent on the test kit for AIDS antibodies. The second concerned the alleged breach of contract by researchers in Gallo's laboratory. The French team claimed that the screening test developed by Gallo's team used the LAV virus which had been supplied to them by the French team. The Pasteur Institute, supported by the French government, claimed that Gallo used Montagnier's virus, LAV, for commercial gain.

The third court case related to the release of scientific documents under the US Freedom of Information Act. It was alleged that documents which indicated that the National Cancer Institute had photographs of LAV in December 1983 were tampered with to delete this information before the documents were supplied to the Pasteur Institute under the Freedom of Information Act (Connor 1987a).

These three cases were settled out of court in March 1987 when the Pasteur Institute and the US DHHS agreed on an 'official' history of the discovery of the AIDS virus.

The two organizations agreed that there should be two patents granted by the US PTO on test kits for AIDS antibodies, one to Gallo's team and one to Montagnier's team, and that each patent should bear the names of both sets of researchers. The agreement stipulated that 80 per cent of the royalties from these patents should go to a new foundation which would spend the money on scientific research into AIDS (Connor 1987c).

The tragedy for AIDS sufferers is that because of the dispute and Robert Gallo's wrongful claim that he discovered the AIDS virus, the AIDS antibody screening test was delayed for over a year during which blood went unscreened, allowing the AIDS virus to infect thousands more people than would otherwise have been the case. AIDS research was retarded because the energies and resources of the scientists involved in this dispute were distracted from the pursuit of AIDS research to that of litigation and administrative proceedings. The outcome for science is that the trust that is essential to the free exchange of scientific information and materials has been jeopardized by the competition for prestige and pecuniary gain.

The imprudence of Martin Cline in his premature attempt at gene therapy in humans illustrates how a desire for prestige can subvert scientific judgement. The UCLA–Hoffman La Roche dispute, the dispute over the AIDS virus and the disputes surrounding the 'billion dollar' Cohen–Boyer patents exemplify how economic incentives have the capacity to undermine the informal traditions and ideals of scientific exchange and communication.

The scientific community has been seen as a perfect moral commonwealth, a self-ruling republic of qualified citizens notable for their *virtu* (see Gibbons and Wittrock 1985). The case studies in biological sciences considered above are an implicit critique of the utopian conception of the scientific endeavour.

J.C. Davis suggests the relationship between science and the utopian ideal is ambivalent and forms one of the central dilemmas of modern Western culture. The techniques of science offer to confer plasticity of the environment upon the utopian designer, as also they offer to reduce contingency to law and so give predictability to design. In return utopia offers social stability and a resource base for scientific activity. But the stability of utopia is jeopardized by the restless and relentless innovations of science. At the same time the utopian impetus towards order and control could threaten the freedom of scientific enquiry (Davis 1984). The utopian ideals of science are incongruous with social reality. The perception of the scientific community as apolitical, fraternal and disinterested—a perception which is often used to dismiss serious criticisms of the integrity of scientific activity—is a myth. The time has surely come when we must finally demythologize the utopian conception of the scientific endeavour.

6 Vermes Chaos: Genetically Engineered Microbes in Biotechnology

Micro-organisms, or animalcules as they were originally called, were first observed under the light microscope by Anton van Leeuwenhoek (1632–1723) of Delft in Holland in the seventeenth century. Until the second half of the nineteenth century, when the nature of their association with putrefaction, fermentation and disease became a focus of microscopy, Leeuwenhoek's discovery of parasites and bacteria remained a curiosity, and those who continued to search for *Vermes chaos*, as Linnaeus (1707–78) classified 'these incredibly small animals', were regarded as eccentrics (Magner 1972). One of the major applications of the new techniques of genetic engineering will be in the engineering of micro-organisms for large-scale applications in biotechnology.

The commercialization of the new techniques of genetic engineering is concomitant with a higher order of risk because as the transition from research to commercial production occurs, so the scale on which genetically manipulated micro-organisms are used will increase. Moreover, large-scale industrial processes are operated by technicians and ancillary personnel who do not possess the same level of knowledge as highly trained and qualified researchers.

Risk-management falls into two parts. The first is an assessment of the hazard posed by the genetically manipulated micro-organism in terms of its survival, multiplication, dissemination, transfer and pathogenic properties. The second is an evaluation of the physical containment measures and industrial safety practices designed to reduce the probability of the escape of genetically manipulated micro-organisms or vectors. This second aspect of risk-management is dependent on the guidelines for large-scale practice, and the regulatory procedures established for their enforcement.

This chapter addresses risk-management policies as they apply to the risks posed to workers and the environment arising from the large-scale use of genetically manipulated micro-organisms in biotechnology.

6.1   REGULATION OF THE COMMERCIAL APPLICATIONS OF GENETICALLY MANIPULATED ORGANISMS

Large-scale recombinant DNA work is defined as work involving production batches of more than ten litres. The 1976 NIH *Guidelines for Research Involving Recombinant DNA Molecules* stipulated that large-scale experiments with recombinant organisms known to make harmful products be prohibited in the USA unless especially sanctioned. In 1979, the NIH-RAC subgroup for large-scale recombinant DNA work was formed, and in 1980 it published physical containment restriction measures and specific recommendations for large-scale work. In its recommendations, the NIH-RAC large-scale subgroup suggested that an *ad hoc* working group be established with the responsibility of advising the NIH-RAC on procedures pertaining to large-scale operations (Landrigan *et al.* 1982). However, with the commercialization of recombinant DNA techniques, it was recognized that there was a need for a co-ordinated framework of regulations, because the existing regulatory scheme of NIH Guidelines was binding only on institutions receiving NIH funds (see 3.3). This left the burgeoning private DNA research industry wholly outside the regulatory framework. Moreover, the NIH-RAC was not designed to cover the diversity of commercial biotechnology and lacked the necessary expertise to evaluate the potential risk of certain applications of recombinant DNA technology.

Throughout the 1980s, under the aegis of the US Office of Science and Technology Policy (OSTP), a co-ordinated framework for the regulation of recombinant DNA technology in biotechnology in the USA was evolved. The co-ordinated framework was first proposed by the OSTP in December 1984. The proposal was that five main regulatory agencies, the NIH, the National Science Foundation (NSF), the United States Department of Agriculture (USDA), the Environmental Protection Agency (EPA) and the Food and Drugs Administration (FDA), should each have a recombinant DNA advisory committee to review and regulate recombinant DNA research proposals under existing laws. Such research proposals would be divided among them as follows: the NIH and the NSF would regulate grant-supported research; the USDA would regulate food-producing biotechnologies; the FDA would regulate drugs; and the EPA would regulate pesticides and chemicals (McCormick 1985). In November 1985 the OSTP published a revision to the 1984 co-ordinated framework and an index of relevant statutes for the regulation of biotechnology appeared in the Federal Register. Finally, the US federal government policy statement for regulating the products of recombinant DNA technology, the Co-ordinated Framework for Regulation of Biotechnology, was published in the Federal Register under the aegis of the OSTP in 1986.

To assist in the process of standardization of regulatory policy, in March 1986 a bill to establish the Biotechnology Science Co-ordinating Committee (BSCC) was introduced in the US Congress (Biotechnology Science Coordination Act of 1986). The BSCC is comprised of senior officials of the federal agencies involved in biotechnology research and product regulation. The charter of the BSCC was to address scientific problems caused by genetically engineered organisms and to establish a Biotechnology Science Research Program to support the research and regulation of the biotechnology sciences including the release of genetically engineered organisms into the environment and the use of such organisms in manufacturing and agricultural activities.

Written into the BSCCs charter was a clause which stated that it was to be disbanded at the end of 1987. In December

1987 it was decided to retain the BSCC with its existing functions. In future, however, it will liaise with a new life sciences committee which is to have a broader membership than the BSCC, including representatives from the Office of Management and Budget, the US Trade Representatives Office and most cabinet-level agencies (Palca 1987b).

The BSCC was created to serve as a co-ordinating forum for addressing scientific problems, sharing information and developing consensus; to promote consistency in the development of federal agencies' review procedures and assessments; to facilitate continuing co-operation among federal agencies on emerging scientific issues that extend beyond those of any one agency; and to develop generic scientific recommendations that can be applied to similar, recurring applications. Its brief was to establish interagency co-ordination of science issues related to research and commercial applications of biotechnology within the Federal Coordinating Council for Science, Engineering and Technology (FCCSET). It receives information regarding the scientific aspects of biotechnology applications submitted to federal research and regulatory agencies for approval and seeks public participation on issues of generic interagency concern. However, the BSCC has no authority to review the regulatory decisions of individual federal agencies for the purpose of approving such decisions.

A key objective of the US federal government policy statement for regulating the products of recombinant DNA technology, the 1986 Co-ordinated Framework for Regulation of Biotechnology, was to resolve the overlapping jurisdictions of the various federal agencies involved in biotechnology regulation. The Co-ordinated Framework set up a matrix establishing the duties and responsibilities among the FDA, EPA and USDA, the three federal agencies involved in regulating biotechnology under present laws. The role of the EPA in regulating the release of genetically engineered bacteria, viruses, fungi, cells and tissue and their use in manufacturing in the USA was clarified in March 1986 by a bill amending the Toxic Substances Control Act (TSCA) (Biotechnology Science Co-ordination Act of 1986). The TSCA is an Act which was developed to regulate the use of chemical products and was therefore not designed to regulate the use of living organisms. The bill amends the TSCA by the addition of a section governing the regulation of the release of genetically engineered organisms into the environment and of the use of such organisms in manufacturing.

Under the amended TSCA, a permit is required from the EPA in order to release genetically engineered organisms into the environment, use a genetically engineered organism in manufacturing or to distribute commercially a genetically engineered organism intended for release into the environment or for use in manufacturing. The Administrator of the EPA was required to establish a permanent Biotechnology Advisory Panel with expertise in risk-assessment and risk-management methodologies related to biotechnology and genetic engineering. The role of the panel is to review and assess applications for all types of permits authorized under this amendment, and make recommendations with regard to which permits should be granted, denied, modified, withdrawn, waived, or expanded. Each such permit issued has an expiration date determined by the Administrator of the EPA, taking into account the recommendations of the Biotechnology Advisory Panel. The applicant or permit holder at all times has the burden of demonstrating that the activities in question will not constitute an unreasonable risk to human health, welfare or the environment. The issue of a permit does not affect the liability of the permit holder for any harm or damage which may result from the release or use of the genetically engineered organism.

As in the USA, in the UK, a number of governmental departments and agencies are involved in the development and enforcement of regulations governing the industrial application of genetic engineering. Responsibility for legislation aimed at satisfying requirements of quality, safety and efficacy of medicines produced by biotechnological means rests with the Department of Health and Social Security (DHSS). The Ministry of Agriculture, Fisheries and Food (MAFF) is responsible for regulations and guidelines covering veterinary, agricultural and food products, and animal and human health, and the Department of the Environment (DoE) is examining appropriate means for the control of releases of genetically manipulated organisms into the general environment. The Health and Safety Executive (HSE), which is the secretariat for the Advisory Committee on Genetic Manipulation (ACGM) (see 3.2 and 3.5) and the Advisory Committee on Dangerous Pathogens, is responsible for occupational health and safety legislation, inspection and enforcement and also offers advice on technical control and risk-assessment in genetic manipulation work.

The GMAG Advisory Notes (see 3.2) which relate to worker safety and health and the ACGM Advisory Notes which superseded them are supplementary to the provision made in the Health and Safety at Work Act (HSW Act). It is not compulsory to follow HSE Advisory Notes because they are intended for use as guidelines to enable work to be conducted within the HSW Act. However, whilst the Advisory Notes themselves are not statutory, non-compliance with them could result in a breach of the HSW Act, which is a civil offence. This makes the UK unique amongst countries with recombinant DNA guidelines in that non-compliance with the Advisory Notes could be an offence under the provisions of the Health and Safety (Genetic Manipulation) Regulations (1978), which are supplementary to the HSW Act.

The Health and Safety (Genetic Manipulation) Regulations (1978) require notification to the HSE of the intention to carry out genetic manipulations as defined in the regulations and the provision of details of individual experiments. However, whereas it is compulsory to notify the HSE of the genetic manipulation of organisms, it is not compulsory to give notification of the use of genetically manipulated organisms in a contained environment or the planned release of such organisms. Thus, although the ACGM issues detailed guidelines indicating how an employer may discharge his or her responsibilities under the HSW Act, the notification scheme for the large-scale use or release of genetically manipulated micro-organisms is not mandatory. Individual proposals submitted under the Health and Safety (Genetic Manipulation) Regulations (1978) and the notification scheme for large-scale work are considered on a case-by-case basis by the HSE and circulated to members of the ACGM, with confidentiality arrangements for commercial information. Where the genetically manipulated organism has warranted only the lowest level of laboratory containment, large-scale work may proceed after notification alone, without awaiting the response from the HSE/ACGM.

In a consultative document published in September 1987, which reviewed the Health and Safety (Genetic Manipulation) Regulations (1978), the Health and Safety Commission (HSC) put forward a proposal to extend Britain's genetic manipulation regulations to cover the use and planned release of genetically manipulated organisms and to ensure effective inspection of establishments undertaking large-scale use. The HSC proposed that voluntary schemes for notification of the contained use or planned release of genetically manipulated organisms in Britain be replaced by compulsory notification and that failure to notify the HSE would become a legal offence (Health and Safety Commission, 1987). In contrast to the USA, the proposed UK legislation does not require notification of the distribution of genetically manipulated organisms.

In the HSCs consultative document, it is proposed that for genetic manipulation work that is categorised as low risk, notification can be filed with the HSE retrospectively in an annual report and for such work it will be sufficient to provide numbers of applications or experiments rather than their details. The proposed legislation, which would make the declaration of genetic manipulation work a legal requirement, is expected to incite industrial opposition because of the case-by-case basis on which the ACGM evaluates and responds to notifications of high risk genetic manipulation work which have to be filed prospectively. Unlike in the USA, where the EPA has 90 days in which to provide evidence that a genetically engineered organism for which a commercial use permit is sought is unsafe for commercial use, the ACGM has no specified time limit. Impatience at the length of time evaluation on a case-by-case basis by the ACGM can take, and frustration by the realization that as the number of notifications increases, so the delay will increase, has increased pressure from UK industrialists for the ACGM to introduce more rationalized assessment procedures, in line with those of the regulatory authorities in the USA (see 6.3) (Newmark 1987b).

There is no worldwide consensus on what the guidelines or codes of practice for large-scale production involving the use of genetically manipulated organisms should be. In Japan, where guidelines for recombinant DNA work apply to government-funded research only, government permission is needed for each large-scale experiment. In the UK, where the genetic manipulation of organisms falls within the scope of legislation on occupational health, the ACGM applies a non-mandatory case-by-case study system for the control of large-scale work, and the contained use and planned release of genetically manipulated organisms has yet to be regulated. In the USA, risk-assessment of recombinant DNA work is subject to a rationalized approach under environmental legislation with certain categories of genetically engineered organisms subject to a lesser review procedure (see 6.3).

An important contrast between the USA and the UK is that in the USA all agencies and firms require a permit from the EPA in order to release, distribute or use genetically engineered organisms whereas no such permit is required in the UK. Furthermore, the UK Guidance Notes of the ACGM and the proposed changes in the Health and Safety (Genetic Manipulation) Regulations (1978) guarantee confidentiality of commercially sensitive information, whereas the amended TSCA in the USA, in accordance with the Freedom of Information Act, requires that public notice be given locally of the proposed use, distribution or release of genetically engineered organisms. Unfortunately, none of the regulatory systems require that genetically manipulated organisms for commercial use carry a 'genetic marker' which, like a computer bar code, would uniquely identify the source of genetically manipulated organisms which became an environmental or human health hazard (see 6.2).

The chemical, pharmaceutical and food industries in the European Economic Community (EEC) are urging that the regulations and the system for the evaluation of the risks of the commercial application of recombinant DNA technology be consistent throughout the EEC. They have proposed to the European Commission a stepped approach to the assessment of the potential risks of the large-scale use recombinant DNA technology and they recommend that each stage, from laboratory experiments through to full commercialization, be considered separately because the impact assessments of potential risks have not yet been well-established. However, were an international code of practice for scaled-up recombinant DNA work to be established, there would also need be an international inspectorate to police safety procedures because of the enormous variation in competence and experience in operating large-scale biotechnology reaction vessels (Bull *et al.* 1982).

6.2 MICROBIAL RISK-ASSESSMENT

Whatever the regulatory guidelines and enforcement procedures, in large-scale biotechnology production, recovery and downstream processing, it is impossible to avoid leakages of microbial agents. The major concern regarding the use of recombinant DNA technology in biotechnological reaction vessels is the hazard to worker health and environmental pollution arising from the failure to contain recombinant micro-organisms. The nature and severity of this hazard depends on the capacity of the modified microbe to cause harm, to survive, to multiply, to disseminate and to transfer its genetic modification. In order to assess the potential of a recombinant microbial strain to cause environmental or health damage, it is necessary to characterize both the natural history of the unmodified organism and the nature of the genetic modification to which it has been subjected for indications of its likely interactions in the biosphere.

The diversity of microbial life has traditionally been a resource for industrial microbiology where it has been common practice to seek useful microbial types in natural settings and then enhance specific properties by *in vivo* mutagenesis and subsequent artificial selection in the laboratory. Accurate identification is a prerequisite for predicting the behaviour of a micro-organism in the environment and for assessing the risk of using it as a host for recombinant DNA. In theory, records of strain history should constitute reliable identification. However, errors in reporting, testing or labelling can occur and cultures can become contaminated. For these reasons, when an application is made to use a recombinant micro-organism in a scaled-up process, it is important for risk-assessment that the 'in-house' strain identity be reconfirmed, preferably by an independent agency, such as the American Type Culture Collection (ATCC) or the National Collection of Type Cultures in London. Independent strain analysis and certification of microbes is, however, impeded by industrial secrecy when manufacturers refuse to disclose information about the nature of manipulation to which the microbe has been subjected (Strauss *et al.* 1986).

Prior to the advent of microgenetic engineering, micro-organisms were identified, described and assigned to taxonomical groups on the basis of observable similarities and differences in the characteristics of the intact micro-organism. The methods of microgenetic engineering allow the inclusion of the characteristics of the genetic material in the identification and description of micro-organisms. This has led to confusion about some previously assigned taxonomical categories as analysis of the DNA has revealed unexpected similarities and differences among some micro-organisms (Strauss *et al.* 1986). In some cases older names and relationships have been abandoned in favour of new names and groupings as new technologies, such as DNA sequencing, provide the basis for new taxonomic criteria based on new types on data.

The use of microgenetic engineering techniques in microbial classification is important in microbial risk-assessment because it may reveal potentially harmful characteristics that would otherwise go unnoticed. Certain genes, for example, may be cryptic—unexpressed legacies of a former evolutionary relationship. Cryptic genes are DNA sequences which code for properties that were important in the history of the species but which are apparently redundant in the present. Genes coding for potentially hazardous properties, such as toxin production, may be present in cryptic forms in apparently innocuous close relatives of pathogenic or toxic micro-organisms. A potential hazard arises from the reactivation of the ancient functions of cryptic genes through mutation (Hall *et al.* 1983). In the absence of microgenetic analysis, there is a high probability that potentially hazardous cryptic genes would not be identified in micro-organisms under review for use in recombinant DNA technology.

The predictive value of any microbial taxonomy is limited by the lack of detailed understanding of the dynamics of evolution in this kingdom. The existing Linnaean system of bacterial taxonomy is ahistorical: it is a poor reflection of the natural (phylogenetic) relationships among bacteria and virtually excludes evolutionary considerations from the study of microbiology (Stackbrandt and Woese 1984). New evidence on the generalized exchange of genes among bacteria substantiates the hypothesis that bacteria are not subject to genetic isolation and thus cannot be considered to be divided into true species (Sonea and Panisset 1980).

The description of a microbial strain should include characterization of extra-chromosomal elements, such as plasmids and phages, which may not be part of either the strain history or the description of the genetic manipulation to which it has been subjected, but which may influence the strain's behaviour. Both phages and plasmids may transport portions of the bacterial chromosome into other organisms. Plasmids, for example, frequently carry drug-resistance genes, and have in many cases been shown to carry traits important for pathogenesis (Timmis and Puhler 1979)

In order to develop an adequate system for the management of risk there is a need for a systematic organization of existing relevant information from currently available sources. The actual amount of information available on a micro-organism is dependent upon the total publications relating to that organism. For some organisms, such as *E. coli*, *Pseudomonas*, *Salmonella* and *Clostridium*, a large amount of information exists, but for other species available information is slight. However, the organisms for which information is available do not represent the limits of biotechnology's horizons, and in some cases bacterial species which have barely been described and whose basic physiological properties have hardly been determined are the subject of recombinant DNA manipulations. Even for those micro-organisms which have been studied extensively, the set of knowledge that has been amassed relates to their behaviour in controlled culture conditions rather than to their behaviour in mixed populations in natural ecosystems (Sharples 1987).

The science of ecology in general is still underdeveloped and at present is more descriptive than predictive. Microbial ecology is even less well developed than ecology in general. It is imperative that microbial ecology is further developed and made a more efficient tool for evaluating ecological risks arising from the escape or deliberate release of recombinant micro-organisms.

6.3 GENETIC ALCHEMY

In addition to a complete description of the natural history and taxonomical relationships of the unmodified micro-organism, risk-assessment should include a thorough description of the organism created by such manipulation. Two basic ways in which the cells of a genetically manipulated organism may differ from its naturally occurring counterpart are that genetic material may have been added to or removed from its genome.

Deletion mutants are cells or organisms from which genetic material has been removed. One application of deletion mutation is to engineer enfeebled microbes which are deficient in certain aspects of their life-cycle. The resultant microbial deletion mutants may be used as enfeebled hosts for recombinant DNA. Another application of deletion mutation is in the creation of mutants which lack the gene considered to be the basis of their pest-status. It is hoped that such deletion mutants will compete with and displace their natural counterparts, acting as biological pest control agents. The 'ice-minus' strains of *Pseudomonas* bacteria created for environmental release for agricultural purposes are examples of such deletion mutants (see 7.1).

Deletion mutants are, in general, considered less of a potential hazard than those organisms into which additional genetic material has been inserted. This view has been taken to its extreme by the Domestic Policy Council (DPC) Working Group on Biotechnology in the USA, in whose opinion a genetically manipulated organism that contains no new genetic material may reasonably be treated like any other altered organism and therefore should be exempted from regulations pertaining to recombinant organisms. In the USA, applications to use gene deletion mutants for commercial purposes are not referred to the EPAs Biotechnology Advisory Panel (see 6.1) for higher review. Similarly, in the proposed revisions to the Health and Safety (Genetic Manipulation) Regulations (1978) in the UK, deletion mutants are not included in the definition of genetic manipulation which means that under the proposed new legislation it might not be mandatory to notify the ACGM of the contained use or planned release of deletion mutants (Health and Safety Commission, 1987; see also 6.1). Although there is worldwide commercial pressure to rationalise the approval procedures for the large-scale use of genetically manipulated organisms, such simplistic genetic rules for exemption may prove disastrously mis-placed. For example, it has been discovered that in the case of certain adenoviruses, a gene deletion in the viral genome is the key factor resulting in its increased virulence.

Those microgenetically manipulated cells or organisms which contain a gene insert may be classified into two categories in terms of risk-assessment; those in which the gene insert is designed to be active within the host cell in order to alter its properties in some way, and those in which the gene insert is designed to be 'silent' or inactive—the gene insert is not designed to alter the function or characteristics of the host cell in any way.

One potential application of deletion mutants and organisms with silent gene insertions is to assess the characteristics of the colonization of the environment by an exogenous strain of organisms. When recaptured, the deletions and gene insertions would act as genetic markers, distinguishing the released organisms from naturally occurring strains. Deletion mutation and silent gene insertion could thus supplement existing methods of assessing the impact of the release of foreign organisms into the environment. The release and tracking of deletion or silent gene insertion mutant strains could be made a pre-requisite for evaluating environmental hazard prior to the release of the same strain containing a gene insert which is designed to be expressed. Deletion or silent genetic markers could also be used to label micro-organisms used in biotechnological production. This would enable the source of micro-organisms which escaped to be traced, and would enable manufacturers to be held responsible for containing micro-organisms used in biotechnology reaction vessels for manufacturing purposes.

The majority of organisms engineered for commercial use will contain an active gene insert which increases their productivity or adds a new metabolic capability. For example, the addition of human insulin genes and gene-control sequences to bacteria to convert them into production units producing more than 20 per cent of their dry weight as the foreign protein, insulin. In the risk-assessment of organisms into which active genes have been inserted, it is necessary to ascertain the function of the gene or its product in its normal setting, from which some degree of risk can be estimated. For example, notwithstanding unforeseen interactions between the host cell genome, the gene insertion sequence and the cloning vector, a gene insert designed to endow the host organism with the ability to secrete a toxin is likely to be considered more hazardous than a gene insert which endows the host organism with the ability to synthesize an amino acid or a pigment it previously lacked.

The US 1986 Co-ordinated Framework sets up a regulatory matrix establishing the duties and responsibilities among several of the federal agencies involved in regulating biotechnology in the USA under existing laws and includes statements of policy from the FDA, the EPA and the USDA describing how they intend to apply their regulatory

authority to biotechnology products. In their policy statements the FDA and the USDA have announced that they
intend to regulate recombinant organisms no differently
from strains obtained by traditional techniques. However,
in its policy statement, the EPA has taken a less lenient
approach to the regulation of recombinant organisms than
the FDA and the USDA. The EPA intends to apply more
stringent regulations to recombinant organisms than to non-
recombinant organisms, but to apply the review procedures
selectively, with certain categories of recombinant organisms
being exempt from federal review, or subjected to a less
rigorous review procedure. In order to differentiate between
different categories of recombinant organism, the EPA has
adopted a strictly genetic definition to evaluate risk and
nominated two categories of organism which should be
subject to federal review. These are 'pathogens' and
'intergeneric organisms'. A pathogen is defined to include
micro-organisms that 'belong to a pathogenic species or
that contain genetic material from source organisms that
are pathogenic'.

A genus (pl. genera) is a taxonomic subgroup in the
Linnaean hierarchical classification system. Under this
hierarchy, all living things are divided into two large groups
called kingdoms, each of which is divided into a series of
major subgroups called phyla (sing. phylum). Each phylum
is further divided into a series of successively smaller groups
known as classes, orders, families, genera and finally
species. There is generally only one kind of organism in
each species. An intergeneric organism is one formed by
the 'deliberate combination of genetic material from sources
in different genera'.

The Industrial Biotechnology Association (IBA) in the
USA has objected to the definition of pathogen in the
EPAs policy statement because it includes organisms that
have inserts of genetic material from known pathogens
regardless of whether the inserts are thought to be involved
in pathenogenic mechanisms. The IBA favours further
exemptions from federal review based on the likelihood of
conferring pathogenicity with the transferred sequences,
and indeed the BSCC (see 6.1) is considering expanding

the categories of genetic modifications that will be exempted from federal review.

Specifically excluded from the definitions of pathogen and intergeneric organism are organisms that are well-characterized and contain only non-coding regulatory regions. Non-coding regulatory regions are not silent gene insertions, but are gene insertions which are intended to affect the level of expression of other genes. Many genetic engineering experiments will seek to change the regulation of one or several genes so that they either produce their products in greater quantity or in response to different environmental signals. Such manipulations may not involve the insertion of genes from organisms of different genera or from pathogens but may none the less constitute a specific hazard if, for example, the recombinant strain produces many times more toxin than the corresponding wild strain. Clearly, the level of expression is an important element in assessing the potential hazard of a construction, and therefore recombinant organisms that are well-charac-terized but contain non-coding regulatory regions should not be exempt from federal review. Furthermore, although the pathogenic effects of recombinant organisms are of great concern, they are not the only detrimental effects that recombinant organisms can wreak on the biosphere. Past experience with non-genetically engineered organisms indicates that the introduction of foreign organisms into ecosystems can have catastrophic effects on indigenous species, through competition and predation (Mooney and Drake 1986).

In the EPA policy statement in the 1986 Co-ordinated Framework and in the opinion of the DPC Working Group only 'new organisms' have potential hazards to which special attention must be paid. 'New organisms' are defined as intergeneric organisms, that is they contain genetic material from organisms of different genera. The EPA maintains that intergeneric organisms are more likely to exhibit new combinations of traits and their interactions in the environment are less predictable. Non-pathogenic organisms that are not intergeneric (i.e. intrageneric organisms) will be exempt from federal approval by the EPA. This regulatory distinction between intrageneric organisms and intergeneric organisms is not compatible with known scientific theory. The assumption underlying this regulatory distinction is that the level of potential risk arising from intergeneric organisms is always less than that arising from intrageneric organisms. There is, however, no scientific evidence to support this assumption; indeed, there are reasons to suppose that the opposite may be true, namely that distantly related organisms are less, rather than more, likely to yield hybrid organisms that are dangerous (Davis 1987). Moreover, applying this distinction to bacteria poses a practical problem because bacteria are not subject to genetic isolation and the classical taxonomic classification does not apply to them. Thus, it would be extremely difficult to ascertain whether two species of bacteria were of the same genus.

It is sometimes argued that the addition of known genes to the genome of an organism currently considered safe in its habitat should be no greater environmental threat than that organism without recombinant genes (e.g. see Brill 1985). However, limiting factors in the environment, such as low nitrogen, low temperature and predation, define the ecological roles of particular species. For example, when the desert is irrigated, many moisture-limited species suddenly appear, and when insect predators are removed, populations of many phytophagous insects are free to multiply to economically harmful levels. Although the environment exerts a directing force on evolution, that force (i.e. natural selection) requires genetic variation. The addition of a well-defined gene insert can provoke unpredictable effects on the physiology, morphology or ecology of the host strain if the genetic modification alters the survivability of the organism with regard to limiting factors in the environment. Slight modifications in only one or a few genes are implicated or clearly documented in many phenomena involving changes in environmentally important phenotypes in all manner of organisms. Examples include expansions of the host ranges of insect and micro-organism pests or parasites and the acquisition of resistance to chemical control agents in insects and bacteria (Halvorson *et al.* 1985).

The effect of the release or escape of genetically manipulated organisms which overcome natural limiting factors is difficult to forecast and cannot be predicted from the nature of the genetic modification alone but can only be investigated through intensive field trials. Moreover, biological containment through genetic enfeeblement provides no grounds for complacency. The extent to which genetic enfeeblement actually reduces fitness is dependent upon the environmental context in which an organism exists (Sharples 1987). That most genetically enfeebled recombinant species will fail to reproduce in the environment is less important than the fact that those few that do succeed may spread rapidly and displace existing species (Wheale 1987).

In addition to unpredictable effects on the habitat and growth properties of the host strain through the addition of a well-defined gene insert, there is the potential for unpredictable effects arising from the transfer of the vector containing novel genes to other organisms in the environment. Although vectors are genetically enfeebled for safety, low rates of transfer of 'safe' plasmids have been observed in the laboratory (Cohen and Laux 1985). Just because it has been demonstrated that a plasmid transfers poorly under laboratory conditions, it cannot be assumed that it will not be transferred to other species in the wild or that it will not be stable in other species (Strauss *et al.* 1986). This problem applies especially to microbes because the extent of gene transfer among them is so extensive that the whole kingdom of procaryotes has access to a common gene pool and is capable of large evolutionary jumps.

The various national regulatory bodies responsible for assessing the risks of proposed large-scale applications of genetically manipulated organisms are under pressure from industry to rationalize their review procedures by creating classes of genetically manipulated organisms which, for genetic reasons, will be exempt from case-by-case assessment. Whilst the characteristics of the unmodified organism and the nature of the genetic modification to which it has been subjected are important factors in the risk-assessment of genetically modified organisms, there are other important factors, and past experience with non-genetically manipulated organisms indicates that the use of limited non-interactive biological criteria to assess risk may not only underestimate ecologically relevant and important factors, but does not assess the risks arising from incompetence, lack of experience, poor facilities, accidents, sabotage and mis-use.

6.4 WORKER HEALTH AND SAFETY

As advances in microgenetic engineering diffuse into biotechnology, the potential exposure of workers to micro-organisms containing recombinant DNA, microbially cloned products and the reagents used in the industrial processing of such products, will increase. Workers involved include professional staff, production workers, line supervisors, maintenance personnel and security staff. The existence, extent and severity of occupational hazards will vary between different sectors of the biotechnology industry depending on the microbial species employed, the genetic modification to which it has been subjected, the product produced, the processing reagents used, the safeguards taken to reduce the risks and the state of health, competence and experience of the operators (Landrigan *et al.* 1982).

The major worker health concern regarding the use of recombinant DNA technology in biotechnology reaction vessels is that workers engaged in industrial applications of recombinant DNA technology are at risk of being colonized or infected by genetically manipulated organisms (Rosenberg and Simon 1979). Physical containment is the primary method of protecting workers in biotechnology from exposure to genetically modified organisms. However, in large-scale biotechnology production, recovery and downstream processing, it is impossible to avoid leakages of microbial agents.

Microbial strains that are used as hosts and the vectors that are used in recombinant DNA technology are supposedly biologically contained. Biological containment involves the use of 'genetic enfeeblement'—deficient microbial strains whose ability to reproduce outside the biotechnology reaction vessel is severely reduced. However, biological containment can never entirely eliminate risk to workers. For example, there is no such thing as an *E. coli* strain which can be guaranteed not to colonize the human intestinal tract under any circumstances due to the possibility that genetically enfeebled strains may revert to robust forms. Furthermore, it has been found in risk-assessment experiments that whilst colonization of the intestinal tract of healthy individuals by genetically enfeebled *E. coli* bearing recombinant DNA plasmids is poor, human and animal test subjects on antibiotic treatment are colonized in large numbers and for long periods of time (Levy *et al.* 1980; Laux *et al.* 1982). The reason for this is that whereas the normal human intestinal tract constitutes a highly competitive ecosystem containing an abundance of microbial flora, antibiotic treatment reduces the level of normal intestinal microbes and thus reduces the number of bacteria with which infecting recombinant bacteria have to compete.

Another concern regarding colonization of the intestinal tract by recombinant plasmid-bearing *E. coli* pertains to the transfer of recombinant plasmids from the original strain to normal intestinal micro-organisms which then replicate and become major components of the intestinal flora. Evidence suggests that whilst this occurs with low frequency in healthy individuals, significant passage of plasmids is observed in individuals undergoing antibiotic therapy (Levine *et al.* 1983). A further alarming observation is that the transfer of recombinant plasmids to normal intestinal *E. coli* strains produces strains which are more efficient colonizers of the intestines, and thus better competitors, than their parents (Cohen and Laux 1985). Individuals taking a course of antibiotics should therefore abstain from working with recombinant DNA organisms for the duration of the treatment and for a short while after. Although genetic enfeeblement reduces the risk, recombinant DNA technology using enfeebled strains is not entirely safe, even for workers who are not, nor who have recently been, taking antibiotics.

The protection of workplace safety in the USA is the function of the US agency of Occupational Safety and Health Administration (OSHA). In areas where significant risks have been clearly identified, for example, industrial applications that employ pathological agents and harmful chemicals, the OSHA can lawfully adopt regulations which will reasonably be expected to protect workers from such dangers. It may not be able to adopt and require compliance with the NIH recombinant DNA Guidelines under the authority of the Occupational Safety and Health Act because the OSHA has to prove the existence of palpable risk before it can impose health standards and the risks of genetic engineering have yet to be proven (Korwek 1981).

The Co-ordinated Framework for the regulation of biotechnology proposed by the OSTP (see 6.1) lists twenty regulatory agencies. However, of these, only five agencies participate in the BSCC and form recombinant DNA advisory committees. The membership of the BSCC includes two representatives from the USDA, one from the FDA and one from the NIH, two from the EPA and one from the NSF. The omission of the OSHA and the National Institute of Occupational Safety and Health (NIOSH) from participation in the BSCC seems to indicate that the US government does not consider workers in biotechnology will be subjected to hazards requiring review by a recombinant DNA advisory committee.

In the UK, occupational health and safety legislation, inspection and enforcement are the responsibility of the HSE. The duties with which the employer is charged under the Health and Safety at Work Act (HSW Act) 1974 are 'the provision and maintenance of a working environment for his employees that is . . . safe without risks to health and adequate as regards facilities and arrangements for their welfare at work'. However, this duty is qualified by the phrase 'so far as is reasonably practicable', and the Act has a clear emphasis on self-regulation.

A stringent code of practice for the control of substances hazardous to health has been produced by the British HSEs

Advisory Committee on Toxic Substances. The Committee panel is comprised of representatives from industry, trade unions and local authorities, as well as 'independent experts'. British employers must follow this code of practice if they are to avoid possible prosecution under the HSW Act. The HSE code of practice provides a framework for assessing the extent of the danger of up to 15,000 chemicals, the levels of potential risk from biological pathogens, and the degree of exposure to which workers may be subjected. Under the new code of practice, employers must keep records of such exposure and give regular medical checks to all employees who have to work with such substances. The regulatory apparatus was designed for two main purposes: to protect workers' health, including that of laboratory workers, and to assure the general public that progress in this field is being properly monitored. However, the code does not address the hazards specific to recombinant organisms.

The ACGM Advisory Notes recommend that a Supervisory Medical Officer (SMO) with experience in public health, infectious disease or occupational medicine is required to conduct health surveillance and reviews to check for particular susceptibilities and to investigate unexplained illness (ACGM/HSE/Note 4). However, the probability that a medical surveillance programme for biotechnology workers will detect illness caused by recombinant organisms or by their products or reagents is low, because neither the nature nor the possible time of onset of any such illness is known, and such illnesses may appear in only a few workers (Landrigan *et al.* 1982). There is a risk that rather than being used to monitor the safety practices of employers, medical surveillance will be used as a means to discriminate against workers who are found to have fallen prey to industrial illnesses.

In 1972 the British Committee on Safety and Health at Work, chaired by Lord Robens, reported that the toll of death, injury and ill health encountered in the workplace was unacceptably high. The Robens Report (1972) considered that apathy was the greatest impediment to improvement in health and safety at work and that reform should be directed towards enhancing effective self-regulation by employers and employees, rather than towards more detailed statutory legislation. This is the intention of the HSW Act of 1974 which has a clear emphasis on self-regulation rather than statutory controls.

The planning and operation of a technological process should be separate from, and independent of, its regulation. Ideally, biotechnology should be policed by an independent factory inspectorate adequately qualified and trained to evaluate health risks. Industrial health science has concentrated on the individual worker and not on the technological environment in which he or she is placed. The tendency has been to attribute blame for the incidence of accidents and ill health at work to the individual employee: the individual is blamed for apathy towards safety, accident proneness, genetic susceptibility and carelessness concerning protective clothing. Although adherence to occupational safety standards, such as the establishment of proper work practices and worker education programmes, is of importance in reducing the likelihood of occupational exposures, the primary defence against occupational exposure to recombinant organisms, their products and reagents is strict physical containment. Clutterbuck (1976) suggests that we should try to re-engineer technology to make it safe. He argues that the wearing of protective clothing should be de-emphasized, as should the use of safety spectacles, ear-muffs and barrier creams, and that instead we should focus on attempts at reducing noise levels, preventing the handling of dangerous chemicals and other substances, and minimizing exposure to bacteria, in order to provide a generally much less hazardous working environment than the one which currently prevails.

6.5 BIOGENETIC WASTE

Biological waste refers to any waste products containing self-replicating biological entities or genetic material which can be transferred to such entities in a functional state. It includes sewage, refuse and effluent from biotechnological processes in which naturally occurring micro-organisms are harnessed to transform feed-stock into commercial products. Biological waste resulting from large-scale biotechnology processes employing genetically modified organisms is a new and largely unregulated pollutant and as a subclass of biological waste is referred to as biogenetic waste.

It is not possible to predict with accuracy what the effects of biogenetic waste on the biosphere will be. Like other exogenous species, genetically manipulated organisms have the potential to wreak deleterious effects on the biosphere by competing with and causing pathogenic effects on indigenous species resulting in their debilitation, displacement or elimination. One specific hazard associated with biogenetic waste is the possibility for genetic exchange with other organisms. The potential of biogenetic waste to harm humans and other species must be considered when instituting its regulation.

For biological waste to be treated effectively, all equipment which comes into contact with it should be autoclaved or steam-sterilized after each use. Autoclaving and steam-sterilizing are methods of killing micro-organisms. In the industrial disposal of conventional biological waste, a margin of between 6 and 7 per cent non-sterility is accepted. If this margin is accepted for the disposal of biogenetic waste, then a substantial quantity of novel organisms will be released into the environment.

Since the relaxation of recombinant DNA guidelines, a large amount of virtually unregulated recombinant DNA work has been conducted in laboratories in countries without modern sewage systems and using many bacterial hosts other than the weakened *E. coli* K12. On the basis that few adverse effects have been documented, it has been argued that the new consensus arrived at in the late 1970s, namely that fears concerning the hazards of recombinant DNA technology were overstated, has in fact been borne out by the experience of the mid-1980s. However, so far no adequate administrative programmes have been developed to analyse at sufficient depth the ecological consequences of the disposal of biogenetic waste.

The apparently safe record of the use of recombinant DNA technology has produced a false sense of security which has influenced the regulations governing the large-scale use of recombinant organisms in biotechnology. In December 1984 the OSTP in the USA stated in its proposed Co-ordinated Framework for the regulation of biotechnology that no new laws were seen as necessary for the regulation of biogenetic waste. The OSTP stated that existing laws for conventional biological waste disposal, such as the Marine Protection Research and Sanctuaries Act, the Clean Air and Clean Water Acts, the Superfund or Comprehensive and Liability Act, and the Resource Conservation and Recovery Act could be extended for use in regulating biogenetic waste.

The Natural Resources Defense Council (NRDC), however, has questioned the adequacy of existing laws in the USA to regulate biogenetic waste. It maintains that the current Clean Air and Clean Water Acts and the Resource Conservation and Recovery Act are inadequate in dealing with the emission or discharge of toxic substances into the environment because they do not make provision for emissions and discharges of recombinant organisms and vectors. Under existing laws such emissions and discharges would be regulated as conventional pollutants and the unique risks which recombinant organisms and vectors pose would not be taken into account. The NRDC maintains that adequate environmental protection would require that the waste discharge from industries using recombinant DNA techniques should be 100 per cent sterile and would place the burden on industry to show why their organism or biological process should be exempted from this rule (Vogel 1985).

There should be a system of qualification and validation steps according to which equipment employed in large-scale operations using genetically manipulated micro-organisms and cells should be tested frequently for contamination and records of these tests and of the operators who use the equipment should be kept. Recombinant DNA technology provides a mechanism whereby the manufacturer responsible for contamination arising from biogenetic waste could be unambiguously identified and held responsible. Recombinant micro-organisms can be engineered to contain a genetic marker which distinguishes them from other strains of the same organism. Application by a manufacturer to use a recombinant micro-organism in biotechnology could be made dependent upon the registration with the regulatory authorities of its distinguishing genetic marker. Such genetic markers would enable the regulatory authorities to detect the release of biogenetic waste, trace its source and impose appropriate sanctions on those culpable.

Proposals for pollution control include public ownership, punitive taxation and legislation aimed at the regulation of pollution levels. Imposing penalties on multinational companies has proved to be expensive and cumbersome, but there is clearly a case for strengthening existing international regulations on pollution control if the existing levels of pollution of the air, sea and land are to be significantly reduced (Rowley 1974). Representative democracies are susceptible to influence by the activities of environmentalist pressure groups, and much depends upon the success of the latter in persuading their governments that further biogenetic waste prevention and control measures are necessary both at the national and international level if we are to avoid probable biosphere catastrophe.

The application of recombinant micro-organisms in biotechnology poses a new order of risks to worker health and the environment. The lack of documented adverse effects following the widespread use of recombinant DNA technology over the past decade is being used in support of the argument that recombinant micro-organisms and their vectors will not be harmful to workers in biotechnology or to the environment. As a result of this attitude a framework for the regulation of biotechnology is evolving which does not take into account the unique hazards posed by recombinant micro-organisms and their vectors. Indeed, existing legislation for the purpose of regulating the production and use of chemicals is being used to regulate the use of living recombinant organisms in biotechnology (see 6.1).

It is important to remember that past experience is not necessarily a reliable guide to the future, and that in the wake of unprecedented industrial accidents, evidence invariably comes to light of warnings that went unheeded.

In devising risk-assessment protocols and containment safeguards for the large-scale use of recombinant organisms, it is important to recognise that the sense of security arising from the experience of recombinant DNA technology is more apparent than real since there has been little organized effort to establish if this activity has resulted in hazardous effects.

Risk-assessment for the contained use and planned release of genetically manipulated organisms should be a multi-disciplinary exercise, which takes into account not only the perceived level of hazard posed by the organism to humans and other species, but also incorporates an evaluation of human and physical factors which increase the likelihood that genetically manipulated organisms will escape from or be deliberately removed from the containment facility. Such broad risk-assessment precludes the exclusion of certain categories of commercial applications of genetically manipulated organisms on the basis of limited biological criteria. Furthermore, to ensure that the risk-assessment procedures are meaningful, there should be an independent inspectorate to monitor the standard of physical containment procedures and industrial practices in operation at large-scale facilities.

As the industrial potential of recombinant DNA technology in biotechnology is realized, there is a case for more extensive regulation and control. Regulatory science policy decisions are dependent upon scientists participating in the political decision-making process. But because risk-management is associated with probabilistic risk-assessment, the lack of disinterestedness and certainty of science has been revealed. The new consensus established by the scientific community and discussed in Chapter 3 above, namely that the risks of recombinant DNA technology have been overestimated, has helped to preserve the authority of science and, in particular, has served as a means to justify reduced public participation in the formulation of regulatory policy.

7 Caducean Crucible: Environmental Releases

At the height of the recombinant DNA debate in the 1970s, concern focused on the accidental escape of recombinant micro-organisms from the laboratory. The main argument for the safety of recombinant DNA work was that the microbes employed in such work were weakened laboratory strains which could never establish themselves in natural ecosystems outside the laboratory. Without recourse to a new debate, we are now entering a new era of commercial exploitation of microgenetic engineering in which the products are genetically manipulated organisms—animals, plants, micro-organisms and live viral vaccines—designed to perform specific tasks in the environment.

This chapter focuses on the hazards and regulation of the deliberate release of organisms which have been genetically engineered not only to survive but indeed to flourish in the biosphere.

7.1 RELEASE OF RECOMBINANT MICRO-ORGANISMS

One of the fastest growing areas of application of microbial labour is in environmental applications for resource-recovery, waste-recycling and environmental control. Many of these applications exploit the extraordinary potential of naturally occurring microbes, for example, those that tolerate high temperatures, high pressures and high levels of radioactivity. As early as 1000 BC copper was recovered from the drainage water of mines in the Mediterranean basin, and this practice was well-established at the Rio Tinto mines in Spain by the eighteenth century. However, it was only relatively recently that scientists realized that bacteria take an active part in this leaching process by converting the copper into a water-soluble form that can be carried off by the leach water. Nowadays bacteria are deliberately exploited to recover copper from billions of tonnes of low-grade ore. In the USA almost 10 per cent of copper production is mined by microbes, and in Canada, India and the USSR microbes perform large-scale leaching of uranium from low-grade ore. Microbial solubilization can also be applied to the leaching of other valuable metals such as cobalt, zinc and lead, and microbes are also used to remove sulphur from high-sulphur coal to reduce sulphur-dioxide pollution when the coal is burned.

Genetic engineering is being applied to create a new order of microbial environmental control and resource-recovery agents. There is commercial interest in genetically modifying microbes to digest oil slicks and recover oil from oil shale and tar sands. Acid-producing bacteria are being engineered to produce and withstand higher acid concentrations to improve their metal-leaching capability. The largest applied use of micro-organisms is in waste-water and sewage treatment, which are early examples of biotechnology. However, toxic chemicals from modern industrial processes and agricultural practices defy break-down by naturally-occurring microbes. Consequently, there is great interest in genetically engineering strains of microbes to detoxify dangerous molecules in waste-treatment and disposal.

Early applications for field trials of recombinant organisms have involved the use of plant-associated microbes in agriculture. In 1982 a research team at the University of California at Berkeley, led by Lindow and Panopoulos and funded by Advanced Genetic Sciences, approached the NIH-RAC regarding a proposed field test on a potato crop of a strain of bacterium called *Pseudomonas syringae* (*P. syringae*) which had been genetically engineered using recombinant DNA techniques.

P. syringae microbes nourish themselves on frost-damaged potato. The majority of naturally occurring *P. syringae* are assisted in this through the possession of a gene, the 'ice-nucleation' gene, which codes for a protein product which promotes the formation of ice crystals on crop plants at temperatures between zero and −7°C. If such bacteria are not present on a plant, ice crystals do not form until temperatures drop below −7°C.

The genetically manipulated ice-minus strains have had the gene that codes for the ice-nucleation protein removed using recombinant DNA techniques. The resulting bacteria are ice-nucleation inactive (INA-) and resemble a minority of naturally occurring *P. syringae* organisms in which this sequence of DNA is absent. The researchers claimed that the replacement of normal *P. syringae* bacteria by the ice-minus (INA-) microbes would enable potato plants to survive at lower temperatures and thereby lengthen the potato growing season.

In the original NIH Guidelines for Research Involving Recombinant DNA Molecules (1976) five classes of experiment were 'not to be initiated at the present time', including experiments involving the 'deliberate release into the environment of any organism containing a recombinant DNA molecule', the possible danger of which was estimated to be so excessive as to be entirely proscribed. The first revision of the Guidelines in 1978 stated that the NIH Director may approve individual waivers of the ban provided the experiment had first been published in the *Federal Register* and reviewed and accepted by the NIH-RAC. In the 1982 revision of the Guidelines the classes of prohibited experiments became 'experiments that require RAC review and NIH and Institutional Biosafety Committee (IBC) approval before initiation'.

The NIH-RAC recommended approval of the Lindow and Panopoulos ice-minus field trial. However, environmentalists were not satisfied that the potential for undesirable ecological consequences of the field trial had been given consideration. They foresaw, for example, that the ice-minus bacteria could compete with bacteria other than the target bacteria or that they could grant frost protection to plant species other than the target crop, thus potentially altering the whole pattern of microbial ecology, plant growth and related insect life with inestimable effects throughout the food web. It was also suggested that the potential of the genetically manipulated bacteria to migrate into the upper atmosphere and alter weather patterns had not been assessed.

Jeremy Rifkin, Director of the Foundation on Economic Trends, together with a coalition of environmentalists and concerned scientists, took the NIH to court and were successful in gaining an injunction to prevent the release of the recombinant bacteria, and the experiments were halted by the Federal Court ruling of District Judge John Sirica. Rifkin charged that the NIH had violated the National Environmental Policy Act (NEPA) by allegedly failing to conduct either of the two kinds of ecological analyses defined by the Act: the environmental assessment and the much more involved analysis, the Environmental Impact Statement (EIS). The NEPA is an Act which covers under federal law any activity in the USA which would significantly affect the environment. The NEPA requires that, in their assessment policy, federal agencies take environmental considerations into account. Products produced using recombinant DNA technology may be able to meet the requirements of the NEPA even when such products are not prepared in accordance with the NIH recombinant DNA Guidelines (Korwek 1980). Under the NEPA any proposal involving federal agency action which would significantly affect the quality of the environment must be accompanied by an EIS. However, all that is required to release an agency from the necessity of filing an EIS is for the manufacturer to provide evidence of no adverse environmental effect. The NIH-RAC claimed to have taken into account questions of human and environmental safety and to have concluded that the proposed field trial would not have significant environmental impact. However, at this time there were no ecologists on the NIH-RAC and their competence to judge environmental safety was questioned.

Until the mid-1980s federal regulation of recombinant DNA technology in the USA was largely confined to the NIH Guidelines for Research Involving Recombinant DNA

Molecules and was the responsibility of the NIH-RAC. The Federal Court ruling prevented the NIH from granting approval for further field tests of novel strains pending the disposal of the law suit brought by Rifkin. In order to circumvent the 'moratorium' on field experiments of genetically engineered organisms, the Monsanto Corporation applied to the Environmental Protection Agency (EPA) for clearance to conduct a field test of its genetically engineered microbial pesticide designed to protect the roots of corn plants against black cutworm. This represented the first application to the EPA to conduct a test of a recombinant microbe in the environment (Sun 1985). At the time of this application the EPA had no clear mandate for controlling the release of recombinant organisms into the environment other than for genetically engineered microbial pesticides for which it was developing regulatory guidelines.

Naturally occurring microbes, including bacteria, fungi, viruses and protozoa, were first registered as pesticides in the USA in the late 1940s and currently there are several hundred applications of biological control of pests. Hazards are posed by the release into the environment of both chemical and microbial pesticides as a result of potentially toxic, allergenic, reproductive, teratogenic or oncogenic effects on the health of non-target species (including humans) and environmental pollution. In addition to these hazards, microbial pesticides may be able to multiply, mutate, reproduce and spread to environments other than those to which they are applied and to exchange genetic information with other related micro-organisms, and they pose the further hazard of being potentially infectious and pathogenic in non-target hosts.

The EPA regulates the use of pesticides in the USA under the authority of the Federal Insecticide, Fungicide and Rodenticide Act (FIFRA). The minimum information and data required for pesticide product registration under the FIFRA focuses on three areas: product analysis—the identification and quantification of all active and inert ingredients; toxicology—the results of short-term studies in several mammalian test species as a guide to potential adverse human effects; and ecological effects—the results of tests of the effects on non-target species. Of major concern in the regulation of genetically engineered microbial pesticides under the FIFRA were potential hazards specific to genetically engineered organisms, which would not be identified by the established testing scheme for non-genetically engineered microbial pesticides. This relates to the stability and specificity of the inserted genetic material, and the potential for exchange of the microgenetically engineered characteristics with other organisms in the environment (Betz *et al.* 1983).

In November 1985, under the FIFRA, the EPA granted permission to Advanced Genetic Sciences (AGS) of Oakland, California, to conduct small-scale field tests of a product which AGS plans to market under the trade name 'Frostban'. Frostban is a mixture of two strains of genetically altered bacteria, *Pseudomonas syringae* and *Pseudomonas fluorescens*, which is designed to prevent frost damage to strawberry blossoms (Boffey 1985). The bacteria are ice-minus (INA-) deletion mutants. It is claimed that the anti-frost microbes will compete with and displace naturally occurring *Pseudomonas* strains from their biological niche, and Frostban has thus been classified as a pesticide.

In attempting to ascertain the ecological effects of the modified bacteria, however, AGS violated federal policies by conducting an unauthorized environmental release experiment. AGS injected the altered bacteria into the bark of about four dozen fruit trees growing on the roof of a building in Oakland *outside* the Company's greenhouse (Schneider 1986a). The EPA fined AGS $13,000 for falsifying scientific data, and temporarily suspended their permit for the experimental release of bacteria onto a strawberry field. AGSs experimental permit for Frostban was, however, reinstated by the EPA in September 1986, and in 1987 Frostban was sprayed onto a strawberry patch in Contra Costa, California (Barinaga 1987a).

By the spring of 1987 there had been four authorized release experiments of genetically manipulated organisms in the UK. One of the approved releases was of a caterpillar virus, another involved nitrogen-fixing bacteria, and the other two were concerned with genetically altered potato plants (Newmark 1987a).

The first UK-authorized release of a recombinant organism was of a microbial pesticide. The release was undertaken in the autumn of 1986 by a team of researchers under the directorship of David Bishop of the Institute of Virology in Oxford. Authority was given for the release after evaluation of laboratory data on the stability and characteristics of the engineered strain and in consultation with the ACGM, the Nature Conservancy Council (NCC), the Ministry of Agriculture, Fisheries and Food (MAFF) and the Department of the Environment (DoE). The experiment is described as a 'monitored release' experiment; recombinant DNA techniques have been used to insert a gene designed to act as a genetic marker. The organism is a baculovirus, a virus which attacks caterpillars of the small mottled willow moth, *Spodoptera exigua* (*S. exigua*). Non-genetically engineered strains of the virus have been used as an insecticide to control caterpillar pests for more than a decade. The gene insert does not code for any new products, and the researchers claim that the slightly altered virus will behave just like its natural parent. The purpose of the gene insert is to distinguish the released virus from naturally-occurring baculovirus, theoretically allowing researchers to identify the engineered virus and its descendants and thus to track the spread of the novel type in the environment.

The genetically marked baculovirus insecticide, named *Autographica californica*, AcNPV, was released in what was described as a 'cabbage patch' ecosystem at an undisclosed location. *S. exigua* caterpillars were infected with AcNPV and then released in a field. It was anticipated that AcNPV viruses would multiply inside and kill the caterpillars.

Containment measures to prevent the dispersal of AcNPV viruses and their proliferation and genetic recombination with other viruses consisted of a net which, it was claimed, would prevent the caterpillars from leaving the field and prevent other insects and predators of the caterpillars from entering the field. However, physical containment measures, such as netting and precautions against soil erosion, can never guarantee 100 per cent containment of engineered organisms.

One advantage claimed for using AcNPV viruses rather than chemical insecticides to control caterpillar pests is their specificity. The AcNPV baculovirus insecticide is believed to have a highly restricted host range among UK moth species, and it is claimed that it will not infect or harm animals or plants other than a limited number of moth caterpillars that are pests. The time taken for AcNPV viruses to kill each *S. exigua* caterpillar is lengthy and is to be shortened by introducing into the virus the capacity to produce a toxin which makes it more harmful to the caterpillar. The long-term objective of this type of experiment is to engineer viral insecticides which, like 'magic bullets', will have a restricted host range, be limited in their ability to persist in the environment and be more virulent. However, a major concern over the use of microbial insecticides, such as AcNPV, is that, like chemical pesticides, micro-organisms genetically engineered for insect pest control will be harmful to non-target species, such as insect species which are economically neutral or economically beneficial (e.g. bees). Another advantage claimed for the use of microbial pesticides is that, unlike chemical pesticides, they will not leave a residue in the environment. However, microbial pesticides, such as AcNPV, are living organisms with the potential to reproduce, mutate, migrate and affect non-target species. Once released, they cannot be recalled, and should they have deleterious effects on non-target species their removal from the environment could prove more problematic than the removal of the residues from chemical pesticides.

Microbial pest control, fertilizer-free crops, more efficient resource-recovery and more efficacious pollution control are some of the potential benefits to be derived from the genetic manipulation of micro-organisms which will be released into the environment to perform their specified role. However, the expansion of the host range for microbial pesticides, the addition of new metabolic capabilities, such as nitrogen-fixation, and the development of microbial strains for pollution control, are genetic modifications which alter some of the major factors limiting microbial growth. Other limiting factors to microbial colonization are the lack of nutrients in the environment and non-nutritional growth regulators, such as acidity, alkalinity, salinity and oxygen concentration. Any manipulation that enhances the survival chances of a micro-organism with respect to one or more of these limiting factors will alter its ecological relationships compared to those of its parental strain.

Populations on the periphery of the biosphere in regions which are dry, cold, hot or saline require the existence of the mainstream biota, and there is mutually beneficial feedback from the periphery to the mainstream. A species that produces ammonia, for example, will stimulate the growth of other bacterial species that can use ammonia as an energy source, and thus alter the community structure. Because micro-organisms actively exchange energy and nutrient elements with one another, a major metabolic change in a genetically engineered microbial species may have effects that extend beyond that species.

In addition to unpredictable effects on the habitat and growth properties of the host strain through the addition of a well-defined gene insert, there is the potential for unpredictable effects arising from the transfer of vectors containing novel genes to other organisms in the environment. The extent of gene transfer among microbes is such that the whole kingdom of procaryotes has access to a common gene pool and is thus capable of large evolutionary jumps. The effect of environmental factors on rates of gene transfer are not well understood. The demonstration that a plasmid transfers poorly under laboratory conditions should not be taken to mean that it will not be transferred to other species in the wild or that it will not be stable in other species (Strauss *et al.* 1986).

Ecosystems consist of a highly complex network of large numbers of mutually dependent living and non-living components undergoing mutually dependent changes in their relative abundances and which process and distribute energy and nutrient elements (Southwick 1976). Microbes are essential to the development of ecosystems. They are the evolutionary precursors of multicellular organisms and act as reservoirs for repopulation after ecological calamity. Microbes play an important role in sustaining ecosystems by recycling non-living components and acting as a buffer to changes in the geosphere. However, ecosystems are in a permanent state of dynamic equilibrium and microbes can play an important role in the dramatic changes which sometimes occur in ecosystems. The ecological consequences of the release of micro-organisms with major novel metabolic capabilities into new environments is therefore very difficult to assess and should be undertaken only with extreme caution.

7.2 THE SECOND GREEN REVOLUTION

Pests and pathogens consume an estimated one-third of the world's crops. Using recombinant DNA technology it is possible to improve crop species by incorporating genes from bacteria, animals and plants, rather than from just related plant species as in conventional plant breeding programmes. Certain herbicide-resistance and pest-resistance traits are governed by single genes, and there is much interest in the prospects of their introduction into plants to make plants resistant to herbicides and pests. Other avenues of commercial interest include the insertion of genes for nitrogen-fixation and tolerance to stress in the environment, such as drought, and increasing the nutritional value of plants by introducing genes which upgrade plant storage proteins for animal consumption.

In principle, the genetic resources of the entire biosphere are now available for crop improvement. The first engineered gene transfer between two plants was achieved in 1981 when USDA scientists created the 'sunbean' by transferring a gene from a french bean seed into a sunflower. Genes from microbes have also been transferred into plants. *Bacillus thuringiensis* (*B. thuringiensis*) is a species of bacterium which produces a toxic substance which is lethal to insect pests. Different strains of the bacterium make toxins with different specificities. *B. thuringiensis* toxin is already in use as a topical insecticide. The Ti plasmid from the plant bacterium *Agrobacterium tumefaciens*, which infects higher plants in some of which it causes a cancerous growth called a crown gall tumour, has been used as a vector to introduce the gene for the *B. thuringiensis* toxin into tobacco plant cells in attempts to engineer plants which synthesize their own insecticide. The regenerated plants express the foreign bacterial genes and produce sufficient insecticidal protein to protect them against damage caused by the larvae of tobacco hornworms. The researchers claim that the gene is stable and is transmitted in accordance with Mendelian principles (Vaeck *et al*. 1987; see also Hilder *et al*. 1987). Recombinant DNA techniques have even been used to transfer the mammalian interferon gene into plant cells using the Ti plasmid vector (Drummond 1983).

The new era of genetic engineering of plants and plant-associated microbes has been heralded as the second Green Revolution, which will provide a technical fix for the problems associated with the first so-called Green Revolution of the 1960s. The first Green Revolution was the replacement of traditional crops by high-yield varieties requiring irrigation systems and expensive inputs of fertilizers and pesticides produced by the world's chemical companies to sustain them.

The greatest fossil energy input into agriculture is that of nitrogen-containing fertilizer. The requirement for chemical fertilizer could be reduced by using recombinant DNA techniques to insert the genes of nitrogen-fixing bacteria into crop plants. Nitrogen is an important component of proteins, which form the bulk of the structures and perform the majority of biochemical functions of living organisms. However, living organisms are just part of the chain of states in which nitrogen exists as it repeats its passages through the nitrogen cycle. When living organisms die, the nitrogen in their bodies is utilized by a variety of bacteria and thereby processed into various nitrogen compounds including ammonia, nitrites, nitrates and nitrogen gas in the air, which comprises 80 per cent of the air by volume. Animals obtain nitrogen by eating plants. Most plants cannot use nitrogen from the air and can only assimilate nitrogen in the form of nitrates. In crop cultivation, the level of nitrates in the soil is elevated by the use of nitrogenous fertilizers.

Certain bacteria and primitive plants are able to tap the reserves of nitrogen in the atmosphere and build it up into protein. This process is called nitrogen-fixation. The best-known nitrogen-fixing organisms are bacteria that live in the roots of leguminous plants such as peas, beans and clover. The bacteria enter the young plant through its root hairs, where they cause a swelling, called a root nodule, in which the bacteria multiply rapidly, fixing atmospheric nitrogen and building it up into proteins. The relationship between nitrogen-fixing bacteria and plants is mutually beneficial: the plant gains some of the bacteria's fixed nitrogen for protein synthesis and the bacteria gain carbo-hydrates for energy from the plant's photosynthetic activi-ties. The insertion of nitrogen-fixing genes from nitrogen-fixing bacteria into crop plants rendering them capable of fixing nitrogen from the air and the use of genetically engineered bacteria with improved nitrogen-fixing ability could reduce the agricultural dependence on petroleum for fertilizers, an increasingly expensive input in the production of food and a cause of pollution in the waterways.

The ecological side-effects of the use of chemical pesticides are often undesirable and difficult to predict. Pesticides that are mercury-compound fungicides have seriously polluted water. A report by the OTA in the USA found that 5 per cent of all pesticides end up in surface water. Wells have been shut down because of farm chemical contamination, and there is growing evidence that the groundwater in many areas of the USA is also contaminated with pesticides (OTA 1984a). Organochlorine compounds, such as DDT, Aldrin and Dieldrin, are persistant, and once applied, their residue remains biologically potent for long periods, thus threatening future crops and wildlife. Chemical pesticides tend to be harmful to beneficial organisms as well as the target species; for example, herbicides disrupt microscopic organisms that allow air and water to move freely through the soil. It is alleged that farm chemicals have been responsible for causing cancer, birth defects and genetic mutations in humans (Manning 1977).

A recent study by the World Resources Institute indicates

that many strains of insect and plant pests have spontaneously developed resistance to agricultural chemicals. The high reproductive capacity of pests facilitates the rapid local growth of populations of mutant pests resistant to pesticides in use in the area. According to the study, at least 150 kinds of fungus and bacteria are pesticide-resistant and about thirty types of weed have developed resistance to the herbicide atrazine (Oka and Pimentel 1976). New pesticides are therefore constantly being developed, an increasingly expensive process.

As a result of the first Green Revolution, there has been greater uniformity among crops and a reduction in the gene pool. This, together with the trend towards monoculture (large areas producing one crop), increases the vulnerability of crops to attack by pests. The very use of herbicides also renders plants more susceptible to attack by insects and plant diseases. Recombinant plants with new genes for resistance to microbial, animal and plant pests and the use of genetically modified microbes as pesticides could reduce the need for spraying crops with chemical pesticides. However, much of the agricultural research effort is being conducted by the agrochemical industry, and initial developments have been largely confined to the transfer of genes conferring tolerance to herbicides rather than undertaking the lengthy and expensive development of pest-resistant plants. The use of herbicide-resistant crops will increase the market for chemical herbicides currently limited in use because they are as deadly to crop plants as they are to weeds.

A number of major agrochemical companies are developing plants with resistance to their brand of herbicide. Ciba-Geigy have funded research at Michigan State University in order to develop soybeans with resistance to the herbicide atrazine. Atrazine, is marketed by Ciba-Geigy under the trade name 'AAtrex', and is widely used on maize. Whereas maize contains enzymes which detoxify atrazine, soybeans do not, and if soybeans are rotated on land previously sprayed with atrazine, the bean crop will be damaged. Researchers have isolated and cloned the atrazine-resistance gene and are attempting to transfer the gene to soybeans and other crop plants. It is estimated that the development of soybeans resistant to atrazine would greatly increase annual sales of atrazine.

Research is also being undertaken to develop crop plants resistant to other herbicides such as DuPont's 'Oust' and 'Glean' and Monsanto's 'Round-Up', which are lethal to most herbaceous plants and thus cannot be applied directly to crops. The successful development and sale of crop plants resistant to brand-name herbicides will result in further economic concentration of the agroindustry market increasing the market power of transnational companies.

There is an intimate relationship between weeds and crops. Existing weeds are likely to benefit from many of the same kinds of genetic novelties, such as herbicide-resistance, pest-resistance, stress-tolerance and nitrogen-fixing ability, that genetic engineers are striving to introduce into crop plants. Genes for these traits may be transferred to neighbouring weeds as a result of naturally-occurring gene transfer between plant species.

It is not necessary to postulate any novel genetic transposition mechanism whereby these new characteristics may be transferred into surrounding plants. Many important weeds are closely related to crops and in many circumstances hybridize freely with crops. The use of recombinant DNA technology in horticulture may be more likely than conventional plant breeding to produce new or more troublesome weeds because recombinant DNA techniques permit the genetic engineer to introduce foreign genes into crop plants. The use of cloning vectors to insert genes generates the hazard borne of the reactivity of these insertion agents within the host cell genomes and their potential to exchange genetic material with, and to infect, cells other than the target cells. The vectors used in inserting novel traits may be able to infect the cells of neighbouring plants, for example, thus endowing them with novel properties (Colwell *et al*. 1985).

The second Green Revolution may ease production and enhance yields in the short-term, but in the long-term it may create health and environmental hazards and undermine agricultural productivity. A pessimistic prediction is that the outcome of the second Green Revolution could be a drastic increase in the requirements for chemical herbicides for use on herbicide-resistant crops developed by agrochemical companies and to overcome the super-weeds that genetic engineers have inadvertently created.

7.3   COLD COMFORT FARM

The new techniques of genetic engineering enable the genetic engineer to mix the genetic material of different species in combinations not possible using conventional cross-breeding. Microgenetic manipulation and cell fusion are conducted on a cellular level and the results are genetically engineered cells. Animals are composed of hundreds of thousands of millions of cells and it would be technically extremely difficult to genetically manipulate all the cells of an animal or even sufficient cells of a target organ in that animal to bring about a desired change. However, when used in conjunction with the new reproductive technologies, the new techniques of genetic engineering can be effective in producing genetically engineered animals. Artifical insemination (AI), has become the leading method of breeding cattle in advanced industrial countries. Egg donation, *in vitro* fertilization (IVF) and surrogacy (see 10.4) enable farmers to maximise the number of offspring of livestock which have genetic traits considered desirable. Embryo duplication enables a single cow to produce twin calves five times a year, and sex selection of six-day embryos enables dairy-breeders to implant female embryos, while beef-farmers can choose to implant male embryos (Merritt 1987).

Experimentation with the use of genetic engineering in conjunction with the new reproductive technologies to produce novel animals has already been undertaken. In 1984, the Institute of Animal Physiology in the UK announced that they had created a chimaera—a novel cell fusion animal that was a hybrid between a sheep and a goat. In 1982, rat growth hormone genes were micro-injected into fertilized mouse eggs. The genetically manipulated eggs were then implanted into surrogate mother mice who gave birth to genetically engineered 'super mice', so-called because the effect of the rat growth hormone was to make them grow abnormally large (J. Williams 1982). In 1987, a refinement of this experiment with more obvious commercial applications was conducted by a team working at the Department of Molecular Embryology at the Institute of Animal Physiology near Cambridge in England which created the first genetically engineered 'big pig'. This was achieved by inserting into the fertilized eggs of pigs a gene control sequence affecting hormones which regulate pig growth. Dr Gilmour, a member of the research team, believes that 'big pigs' could be on the farm by 1992. It is claimed that their advantage over normal pigs would be that they would reach the size of normal pigs at an earlier age and if killed at this age, their meat would be more tender (Hadlington 1987). Poultry are also targets for microgenetic engineering with hormone genes. Genetic researchers have identified a chicken growth hormone gene that could significantly reduce the time taken to rear broiler fowl.

A pioneering experiment in 1987 again used mice as a model system for techniques with commercial applications for livestock. Scientists working at the Institute of Animal Physiology and Genetics Research in Edinburgh demonstrated that it is possible to genetically manipulate mice so that they secrete a sheep protein in their milk. Milk contains a large amount of protein, most of which consists of a few major types synthesized in the mammary gland. In the experiment, sheep genes were micro-injected into fertilized mouse eggs which were then implanted into female mice (Simons *et al.* 1987). Some scientists hope to use these techniques to engineer dairy animals to secrete foreign proteins of high value in their milk.

To date, the use of microgenetic manipulation techniques and cell fusion to genetically engineer animals has not been a commercial success. The sheep-goat chimaera, for example, is a sterile, frail and disabled creature. The success rate of micro-injection of foreign genes into fertilized eggs is exceedingly low. For instance, in an experiment undertaken in 1987 in which 320 mouse eggs were each injected with 200 copies of a foreign gene, only one mouse was subsequently born that was able to make the foreign protein (see 9.1). Many eggs fail to develop properly because the insertion of the foreign genes into the mouse egg genome disrupts, often fatally, the normal development of the eggs. Even in those cases where the insertion itself is not detrimental to the development of the animals, the expression of the foreign gene in uncontrolled quantities and in inappropriate tissues has deleterious effects upon the genetically manipulated animals. For example, a 'fast-growing' pig produced by the USDA with the growth hormone gene of a cow is grossly deformed, and even before reaching the age of two years is crippled by arthritis. It is a boar with extremely short legs, crossed eyes and has a strange wrinkly rust-coloured skin. However, it is capable of breeding and its meat is low in fat. Indeed, the fast-growing pig' produces so little fat that it is at risk of dying from the cold.

It is the technical difficulties of genetically engineering commercially viable livestock (rather than animal welfare reasons) which have determined that the first products of the new genetic engineering to be used for farm animals are veterinary medicines synthesized by genetically engineered bacteria. The veterinary medicines approach has been used to increase milk production in dairy cows. Since 1985 field trials have been conducted in the UK and the USA of several genetically engineered versions of bovine growth hormone, collectively known as bovine somatotropin (BST). The BSTs are chemicaly similar to a growth hormone which occurs naturally in cows. They are produced by inserting a modified version of the gene coding for naturally occurring bovine growth hormone into bacteria. These microbes then produce a version of bovine growth hormone called BST. It is claimed that when microbially-cloned BST is injected into dairy cows, it increases their milk yield without raising feed costs.

In the UK, the manufacture and licensing of medicines

is controlled by the Medicines Act of 1968 under which all veterinary and human medicines are judged on a case-by-case basis according to criteria of safety, efficacy and quality. The MAFF has responsibility for the control of veterinary medicines, and applications to market veterinary products are assessed by the Veterinary Products Committee (VPC), a statutory body of independent experts, which takes advice from the Central Veterinary Laboratory of the MAFF licensing department. Even though the long-term effects of BST on humans and cows are not known, the MAFF issued test certificates under the Medicines Act (1968) to several major drug companies to conduct field trials of the hormone BST with 1,000 British cows. It is claimed that microbially cloned BST is identical to natural bovine growth hormone and that its use as a milk-boosting drug in large quantities poses no threat to human health.

The resultant hormone-boosted milk is sold through the Milk Marketing Board in the UK to consumers who are not in a position to discriminate against it, although consumer surveys conducted in the USA and the UK suggest that consumers would boycott BST treated milk. When assessing veterinary products under the UK Medicines Act (1968) the criterion of safety is applied to safety of the animal, safety to the consumer of the animal, safety to the farm worker and safety to the environment. However, there is some evidence that the use of BST in large quantities may cause adverse effects to the cow, including possible breakdown of the mammary gland. Data relating to experiments on the effect on cattle of injection of BST, which were submitted to the MAFF in support of applications for animal test certificates, were not disclosed on the basis that the Medicines Act (1968) prohibits the disclosure of such information.

The recent interest in growth hormones as drugs to increase the size of livestock is not the first venture of hormone therapy into animal husbandry. Hormones are a class of chemicals produced by animals to perform a co-ordinating role within the body through a general or specific stimulatory or inhibitory effect on biological processes.

Steroid hormones, a category of hormones produced by an animals' ovaries or testes, are used in animal husbandry to fatten farm animals. When administered as a drug, steroid hormones not only affect an animal's sexual development and behaviour, but also boost its growth. Such hormones are used by some bodybuilders to assist them to increase their muscle bulk. The administration of hormones to livestock and poultry for human consumption has the potential to be harmful to human health. In 1980, it was discovered that one of the steroid hormones, called diethylstilboestrol (DES), that was being administered to calves to enhance their growth, contaminated baby food in Italy by way of cheaper cuts of veal causing infant cancers and the onset of secondary sexual characteristics, such as menstruation and breast development, in young children.

From an economic viewpoint, it is difficult to justify the use of BST. BST is used because it is claimed that it increases milk yield by 15-20 per cent without requiring additional feed. However, dairy farmers in the EEC and in the USA produce a surplus of milk. Indeed, surplus milk production in the EEC, which in 1987 was estimated at 17 per cent, represents an economic cost to the EEC, both in terms of storage of the surplus produce and in subsidies to farmers who are paid to reduce milk production.

Milk output in the EEC is controlled by quotas backed by penal levies. For this reason, the only economic way for farmers in the EEC to use BST is to reduce the size of their herd and benefit from reduced cattle feed cost enabling them to reduce milk production costs by an estimated 10 per cent (Merritt 1987). However, these changes will have an adverse effect on the agrofeed industry, the labour market and possibly also land prices, which in the longer term, will place new strains on the farm subsidy systems. If milk production in the EEC and the USA is not restrained and the surplus is placed on the world market at reduced prices it will have disastrous consequences for farmers in developing countries who will not be able to benefit from the use of BST. This is because BST can only be used on high-bred varieties of cattle which need to be fed on expensive concentrates. BST also causes the cows to overheat and therefore cannot be used in hot countries. For these reasons BST will not improve milk production in developing countries.

The utopian vision of the future of agriculture projected by the genetic engineering industry may be summed up by quoting the advertising publicity of Monsanto, a leading US agrochemical company: 'Plants will be given the built-in ability to fend off insects and disease, and to resist stress. Animals will be born vaccinated. Pigs will grow faster and produce leaner meat. Cows will produce milk more economically. And, food crops will be more nutritious and easier to process. And, because most of these products don't require high technology farming practices, they can be used in the agriculture of the Third World, where starvation is a daily event'. However, from the perspectives of animal welfare, human health, ecology and the global economics of food production, the social costs of innovations such as BST, 'big pigs' and 'fast-growing pigs', and pesticide-resistant crops may well exceed the social benefits.

7.4 VIRAL VACCINES

The first vaccination was performed in 1796 by Edward Jenner (1749–1823) using living cowpox virus as a live vaccine against smallpox. Although the vaccines for smallpox and poliomyelitis have nearly eliminated these diseases globally, it has not yet been possible to produce safe and fully effective vaccines against gonorrhoea, syphilis, typhoid, cholera, malaria and AIDS. Recombinant DNA technology has the potential to provide a method for the rapid development of vaccines. Some of these new vaccines may be less hazardous than conventional vaccines. However, others will be living genetically engineered viruses, the use of which constitutes an environmental release.

The main purpose of the immune system is to protect the body from infection in its various forms. The first line of defence against infection is the non-specific immune system, which includes the physical barrier of skin, sneezing, sweat, wax in the ears, interferons and the normal range of microbes which inhabit the skin, the mouth, the throat and the intestines, and which comprise part of the ecosystem of the body. Microbes confer protection from harmful parasites by preventing the colonization of the body by micro-organisms which are pathogenic. If successful, the non-specific immune system halts potential pathogens before they establish an overt infection.

If this first line of defence against infection is breached, the specific immune system can be stimulated to make a very specific immune response. Specificity and memory, two of the key elements of the specific immune system, are exploited in vaccination. Specific immune responses offer no immediate protection on the first occasion the body meets a particular pathogen but they are effective on second and subsequent exposures. The action of the specific immune system enables the body to recover from the disease, and a specific immunological memory is established. The body is said to be immunized against that pathogen so that on subsequent re-infection with the same agent, the specific immune system mounts a much stronger response and no disease results. Second attacks of diseases which stimulate a specific immune response, for example measles and diphtheria, are rare (Kirkwood and Lewis 1983; Roitt *et al*. 1985).

Immunization is the result of the interaction of certain molecules called antigens with cells of the immune system that produce antibodies. An antigen is a foreign substance which is capable of stimulating the body's specific immune system to make a specific immune response leading to their destruction. Generally, the most potent antigens are proteins or large sugar molecules. Molecules on the surface of invading micro-organisms or molecules produced by them may be identified by the body as antigens to which the body responds by mounting an immune response to protect the body from the infection. In the first stage of the specific immune response, antibody-producing cells become activated to make many identical (a clone of) antibodies specific for the antigen. These antibodies attach themselves and label the invading agent by binding tightly to the antigen. Once the foreign agent is labelled by the antibody, scavenger cells of the immune system destroy it.

Vaccines stimulate the production of antibodies. The aim of vaccination is to provoke a specific immune response to various pathogenic micro-organisms so as to confer future long-term protection against their harmful effects. The principle behind vaccination is to expose the subject to antigens that resemble the pathogenic agent sufficiently to activate an immune response in the subject which will protect him or her upon future infection by the pathogenic agent, but which are deficient in their ability to spread infection and cause disease. A safe vaccine is antigenic without being pathogenic.

The safest vaccines for use against viral infections are called synthetic vaccines. Synthetic vaccines are short protein chains comprised of roughly twenty amino acids synthesized in the laboratory to mimic a small antigenic region of a virus's outer coat which can be used as a vaccine against the virus. The laboratory synthesis of a synthetic vaccine against diphtheria has already been achieved by automatically linking amino acids.

Recombinant DNA techniques have great potential in the synthesis of synthetic vaccines. Once an antigenic protein is identified, then its gene can be isolated or constructed and then inserted into suitable host cells. There it can be cloned and expressed as antigenic protein which can be harvested for use as vaccine. In this way, harmless cells or microbes in culture can be used to synthesize antigenic proteins which can be used as synthetic vaccines against harmful viruses. Synthetic vaccines avoid the necessity of culturing viruses harmful to humans for vaccine production and of using human viruses or human viral extracts as vaccines.

David Bishop of the Institute of Virology in Oxford has suggested that recombinant DNA technology be used to develop a synthetic vaccine against infection by the human immunodeficiency virus, HIV, which is responsible for AIDS. He has proposed that antigens from the coat of HIV selected for use as a vaccine against AIDS could be cloned in living caterpillars. Antigen genes would be excised from HIV and spliced into a virus that only infects caterpillars. The caterpillar virus would be used as a vector to insert the HIV antigen genes into caterpillar cells, where it would be cloned. Dr Bishop views the use of caterpillars as small factories for the production of HIV antigen as a 'low-tech' solution to the problem of supplying vaccine to developing countries. Another advantage of the recombinant caterpillar method of HIV antigen production is that the caterpillar virus is unable to infect mammalian hosts and therefore is not believed to be as hazardous to humans as the culturing for vaccine production of live HIV, which are able to infect mammals.

The first commercial genetically engineered synthetic vaccine was a veterinary vaccine designed by Cetus for scours in piglets. In the USA it takes just two years to get market approval for new veterinary vaccines, compared with five to seven years for human vaccines. The relative greater speed with which a commercial license is granted for a livestock vaccine over a vaccine for human use is a stimulus to the development of vaccines for livestock healthcare rather than human healthcare.

The first genetically engineered synthetic vaccine for human use was approved by the FDA in the USA in July 1986. The vaccine, called Recombivax HB, is against hepatitis-B. It was developed by Chiron Corporation of Emeryville, California, and is marketed by Merck, Sharp & Dohme of West Point, Pennsylvania. The new vaccine is produced by cloning an antigen of hepatitis-B virus in a special type of brewers' yeast. Conventional vaccine production is a high-cost, low-volume production system. The synthetic hepatitis-B vaccine theoretically should be much cheaper to produce, but initially was marketed at about the same price as the conventionally produced vaccine.

There are technical problems with the production of synthetic vaccines using recombinant DNA technology. In principle, rapid DNA and RNA sequencing of the genome of a virus would result in the identification of regions of its genome likely to encode proteins suitable for use as synthetic vaccines. However, current theoretical knowledge of protein engineering is inadequate for predicting how a given protein will perform as an antigen from its structure as deduced from its genetic sequence, and in the near future the selection of antigen proteins for use as synthetic vaccines is likely to be by trial and error (Pain 1987). The antigenic properties of cloned proteins will be tested by injecting them into laboratory animals to discover if they immunize the animal against the virus. As knowledge of the relationship between genetic sequence and protein structure, and of the interaction between antigens and the immune system increases, scientists hope to use computer programmes to aid in selecting genes which code for viral proteins suitable for use as synthetic vaccines.

Before genetically engineered synthetic vaccines can become an alternative to conventional vaccines, ways must be found to increase the magnitude and duration of the immunity they confer. Synthetic vaccines do not replicate inside the subject and an adjuvant, a substance injected with antigens which enhances the immune response to the antigens, must be added to them. A new generation of adjuvants and a deeper understanding of the nature of the specific immune response is required before it will be possible to elicit from genetically engineered synthetic vaccines the range of immune responses necessary for clinical immunity.

In synthetic vaccines, recombinant DNA technology has the potential to produce safer vaccines. However, the technical hurdles to the production of synthetic vaccines using recombinant DNA techniques mean that, in the short-term at least, the new techniques of genetic engineering are more likely to be used to create new vaccines involving the use of harmful viruses in their preparation, and possibly in the final product.

Vaccinia, the cowpox virus used by Edward Jenner as a vaccine against smallpox, is a naturally attenuated living virus. An attenuated virus is one whose virulence in human beings has been reduced by mutation. Live viral vaccines, which are attenuated living viruses, multiply in the patient and produce a continuing stimulus to the immune system.

Smallpox, rubella, measles, polio (oral Sabin), yellow fever, mumps and BCG (tuberculosis) are examples of live vaccines. Recombinant DNA technology can be used to engineer living attenuated viruses for use a live vaccines. For example, site-specific mutagenesis can be used to delete portions of the viral genome to create attenuated viruses.

Many researchers consider that it is too risky to use attenuated lethal viruses as live vaccines. Some researchers believe that one way of circumventing this risk is to use hybrid viruses as vaccines. Recombinant DNA technology can be used to add the gene for an antigen of a lethal virus to the genome of a harmless virus, in an attempt to create a harmless living hybrid virus which if used as a vaccine provides immunity against the lethal virus. This principle has been applied in a number of research projects utilizing recombinant DNA technology in the hope of providing vaccines against HIV, the virus responsible for AIDS. Approaches include hybrids between HIV and vaccinia, the cowpox virus, and hybrids between HIV and a relatively harmless virus called adenovirus. The idea is that the living hybrid viruses can be used as live vaccines against HIV infection. However, Professor William Jarrett of Glasgow University, a leading researcher in vaccine research, has severe apprehensions about his former plans to engineer a hybrid HIV–adenovirus vaccine because he now believes that there is alarming evidence which indicates that such an approach to vaccine development is potentially extremely hazardous. The deletion of portions of the adenovirus genome in order to insert HIV antigen genes creates a strain of adenovirus which has proved to be more virulent than its unadulterated parent strain. This is an instance which illustrates that deletion mutants cannot be assumed to be less harmful than their parent strains. Clearly this calls into question the regulatory decision in the USA that gene deletion mutants should no longer be referred for higher review (see 6.3).

Any method of vaccine preparation which involves culturing harmful viruses poses hazards to the subject, to laboratory workers and to the general public. Viruses can only replicate when inside living cells and therefore supplies of virus for use in producing vaccines must be grown inside living cells, either in infected animals or in cells in culture. The cells may contain undetected substances, notably other viruses, that contaminate the vaccine and are transmitted to patients. Any facility where viruses are grown to make a vaccine is a reservoir from which an agent of disease can accidentally be disseminated. Although laboratories handling pathogenic micro-organisms are designated high-level physical containment laboratories, physical measures cannot be relied upon to provide total containment, indeed several people have been infected with harmful viruses as a result of incidents in laboratories handling dangerous pathogens (see 3.6).

The use of living viruses as live vaccines is an environmental release. However, it is not covered by the definition of release in the proposed notification legislation in the UK (see 7.5). The Advisory Notes for risk-assessment and for the notification of proposals for planned releases only apply to genetically manipulated organisms for agricultural and environmental purposes (ACGM/HSE/Note3). The licensing of genetically manipulated organisms developed for human or veterinary prophylactic or therapeutic purposes in the UK is the responsibility of the DHSS and the MAFF, as advised by the Committee on Safety of Medicines (CSM). In the USA, live viral vaccines that are deletion mutants are exempt from regulation by the EPA, the agency now charged with the task of regulating the commercial use of recombinant organisms, and live viral vaccines that are not covered by the definition of pathogen or intergeneric organism are not subjected to review by the Biotechnology Advisory Panel (see 6.1).

Vaccines, therapeutic sera, toxins and antitoxins are classified as 'biologicals' for the purpose of regulation. Biologicals are substances whose purity or potency cannot be adequately tested by chemical means. Safety approval in the USA should be made by the FDA Office of Biologics using unpublished guidelines or 'regulatory memoranda'. In the UK biologicals are licensed on a batch release system evaluated by the National Institute for Biological Standards and Control (NIBSC) which advises manufacturers proposing to submit biological products for the approval of the CSM. Sweden, France and West Germany have guidelines for the production of biologicals and, though no specific legislation on biologicals has been approved, the European Economic Community (EEC) drug regulation aims to harmonise testing and approval procedures of member states. The World Health Organisation (WHO) also publishes guidelines on biologicals.

The first commercial licence for the release of a living recombinant DNA virus was issued on 16 January 1986, not by the FDA but by the USDA biologics licensing division to Biologics Corporation, a company in Nebraska, for a live vaccine called 'Omnivac'. Omnivac is used to vaccinate swine against a form of herpes virus called *Pseudorabies* virus, which causes a disease also called pseudorabies. Pseudorabies produces skin lesions, paralysis and coma in infected pigs, cattle and sheep, leading to death within days of onset. Omnivac is a deletion mutant of the *Pseudorabies* virus, created by the specific deletion of a gene of the virus using recombinant DNA techniques. The engineered virus lacks the gene for thymidine kinase, which is apparently responsible for the pathogenic effect of the wild-type virus and which allows the virus to 'hide' in latent form in the animal's central nervous system. The marketing licence granted by the USDA for Omnivac followed the undertaking in autumn 1985 of field trials in three states on permits issued by the USDA in April 1985.

The Foundation on Economic Trends charged that decisions made by the USDA involving Omnivac violated federal procedures. The USDA and other federal agencies have agreed to comply with the NIH Guidelines of 1978 which require that prior to consenting to applications to make planned releases of genetically engineered organisms the agencies make a complete assessment of their effects and seek counsel from experts from other agencies and from outside the government. Although the Agricultural Recombinant DNA Research Committee (USDA-ARRC) was set up in 1976 to assess and co-ordinate proposals for recombinant DNA in agriculture, the biotechnology regulatory programme at the USDA was dealt with mainly by the Animal and Plant Health Inspection Service (USDA-APHIS), the division which allowed Omnivac to be tested, with assistance from the Agricultural Research Service. Charged with having made its decision without consulting experts in its own agency, the USDA-ARRC, and without notifying other agencies, the USDA was urged to revoke the licence.

In April 1986 the General Accounting Office (GAO), the investigative arm of Congress, charged that the USDA had not formulated a well-defined regulatory structure with regard to deliberate releases of genetically engineered organisms into the environment. The Chief Science Policy Administrator of the USDA, Orville G. Bentley, Assistant Secretary for Science and Education, admitted that the USDA had acted in error in approving the first field test of Omnivac, stating that the application to field test and market the vaccine should have been brought before the USDA-ARRC, whose opinion should have been sought. The USDA-ARRC was later disbanded and an Agricultural Biotechnology Recombinant DNA Advisory Committee was established with the responsibility of reviewing all recombinant DNA research projects submitted to the USDA.

It is alleged that commercial pressure played a part in the decision by the USDA-APHIS to approve licences to market Omnivac. In Minnesota the pig business is a $2.5 billion a year industry involving 20,000 herds. In the four months prior to the disclosure of the licence, Omnivac amassed sales of $250,000. Biologics Corporation, the company that manufactures Omnivac, is said to have failed to submit all of the scientific data that characterized the organism in order to protect its application for a patent on the vaccine. With the approval of the USDA, Biologics Corporation did not notify state officials in Minnesota, Michigan and Illinois—the states in which Omnivac was field tested—Omnivac was a living recombinant virus until after the patent was issued, which was after the field trials had taken place. In response to these criticisms the USDA responded that they did not want to 'impose cumbersome regulations' that might stifle industry (Schneider 1986b).

The USDA defended their decision to permit field trials of Omnivac on the basis that it is a deletion mutant and that it contains no new genetic material. The USDA claimed that it had not violated federal policies in permitting the field trials and granting the marketing licenses for Omnivac since a similar mutant virus created using classical methods of genetic engineering had been approved and used as a live pseudorabies vaccine in the past (Beardsley 1986). David Espereth, chief veterinarian in the veterinary biologics staff at the USDA-APHIS, the division of the USDA which allowed the vaccine to be tested, maintained that the host range and virulence of the virus were unchanged because the recombinant DNA genetic engineering had involved a deletion, not the introduction of a foreign gene. This decision is consistent with the opinion of the Domestic Policy Council (DPC) Working Group on Biotechnology that genetically manipulated organisms that contain no new genetic material, such as Omnivac and the ice-minus strains of frost-control bacteria, can be treated in the same way as any other altered vaccine or organism that has been engineered without using recombinant DNA techniques.

David Espereth also defended the action of the USDA with regard to Omnivac on the grounds that vaccination does not constitute 'release' into the environment. In the view both of the Industrial Biotechnology Association and the Association of Biotechnology Companies, there is no consensus within the biotechnology industry or among the regulatory agencies in the USA of what constitutes environmental release or containment (K. Wright 1986).

As a result of regulatory constraints and confusions, delay, public opposition and unfavourable publicity, genetic engineering companies have begun to conduct their release experiments involving recombinant viruses engineered for use as vaccines in countries where the obstacles appear to be fewer due to laxer legislation and low public awareness. In New Zealand in April 1986, with the consent and funding of the USDA and the permission the New Zealand Ministry of Agriculture and Fisheries and the Ministry of Science and Technology, thirty-seven calves, sixteen chickens and four sheep were innoculated with a live recombinant vaccine at a research station operated by New Zealand's agricultural ministry near Wellington. The vaccine, constructed by Alvin Smith and colleagues at Oregon State University in the USA was a hybrid between vaccinia virus and a gene for an antigen from the sindbis virus, a virus which is transmitted by insects and produces influenza-like symptoms.

In July 1986, a live recombinant vaccine designed to fight bovine rabies, a serious health problem besetting the Argentinian cattle industry, was field tested at a remote research station operated by the PAHO, a United Nations organization dedicated to the study of animal diseases, in Azul about 250km south of Buenos Aires in Brazil. The PAHO volunteered its station for testing the vaccine in the field without informing the governments of Argentina or the USA. The vaccine tested is a hybrid virus constructed using recombinant DNA techniques to splice a gene that codes for an antigen of the bovine rabies virus into vaccinia virus, which was then used as a live vaccine.[1]

Both the sindbis vaccine experiment in New Zealand and the bovine rabies vaccine experiment in Argentina involved the use of living hybrid vaccinia virus as a live vaccine. Two hundred years ago, vaccinia was used to make the first vaccine against smallpox. Now it is a popular choice of host for live hybrid vaccines against infection with harmful viruses, including HIV. Alarmingly, there is evidence that close relatives of vaccinia exist in many parts of the world, for example camel pox in Africa and Asia, cowpox in European cats and cows, monkeypox in Africa and racoonpox in North America. Yet there have been no assessments of whether animals innoculated with the mutant vaccinia could pass it on to animals harbouring a related virus (Joyce 1986).

In July 1987 an Indo–US agreement was signed to enable genetically engineered vaccines developed in US laboratories to be tested in India. The agreement is for a Vaccine Action Programme lasting five years designed to develop and test vaccines and diagnostic techniques for major communicable diseases. Among the vaccines to be tested are a recombinant DNA vaccine against hepatitis-B and the vaccinia–rabies hybrid vaccine used in the experiment on cattle in Argentina. The official motive for the export of vaccines developed in the USA to India for clinical testing is that their efficacy can best be evaluated in places where the diseases they are designed to prevent are most prevalent. However, some of the vaccines covered in the Vaccine Action Programme might for regulatory reasons be difficult to test in the USA, and the director of the Indian Council of Medical Research, fearful that Indian people are to be used as guinea-pigs, is demanding that the vaccines gain approval of the US FDA before they are used in India (Jayaraman 1987).

Large sums are being spent on medical research in North America and Europe to develop vaccines to prevent infection with HIV or to prevent HIV infection from progressing into the disease, AIDS.

The WHO had hoped to establish ground-rules for conducting tests of AIDS vaccines before clinical trials began. However, at the end of 1986 a vaccine to combat AIDS was reported to have been undergoing trials in Zaire since September of that year. The investigator was Daniel Zagury of the University of Paris. The vaccine was reported to be a hybrid vaccine derived from vaccinia recombined with a gene coding for a protein on the surface of the HIV virus. The trials in Zaire were aimed at keeping individuals already infected with HIV from developing AIDS itself. In testing the vaccine's effectiveness, the investigators by-passed the usual preliminary safety tests.

The use of living viruses as live viral vaccines constitutes an environmental release. As a result of the use of recombinant DNA technology, more novel live vaccines will be developed, undergo clinical trials and be made commercially available in the near future than ever before.

Whether hybrid viruses or attenuated viruses, the use of living viruses as live vaccines poses potential hazards to patients, laboratory workers and the general public. Live viral vaccines are living viruses which are capable of mutation, possibly reverting to a virulent form, hence there is a grave risk that the use of live vaccines could instigate an epidemic of the very diseases they are designed to prevent. Even in the absence of mutation, live vaccines may possibly have some unknown long-term effects comparable to the slow-acting viruses responsible for scrapie in sheep. Another possible hazard is that living viruses used as live vaccines will exchange genetic material with other viruses creating novel pathogens. Live viral vaccines multiply in the subject and produce a continuing stimulus to the immune system and can be dangerous to subjects who are immunodeficient, such as AIDS patients, who can be overcome by infection even by so-called harmless microorganisms. In the USA, this knowledge has led to a recommendation that doctors should not give polio or measles vaccines to people suffering from AIDS lest they succumb to infection and illness caused by the vaccine itself.

We urge that a stringent regulatory framework controlling the introduction of new live viral vaccines be devised and that scientists resist the well-established practice of exporting novel medicines and vaccines for clinical testing to countries where legislative obstacles appear to be fewer, and that they take on the responsibility to uphold the same uniformly high standards of safety throughout the world.

7.5 RISK-MANAGEMENT

The application of microgenetic techniques in order to improve crops and livestock, control pests, recover natural resources or degrade pollutants, and produce live vaccines requires the release of living genetically engineered organisms.

The applications for the release of ice-minus *P. syringae* (see 7.1) highlighted the uncertainty surrounding the regulatory framework in the USA for the planned release of recombinant organisms. There was a boundary problem created by the delimitations of the terms of reference between the NIH and other regulatory agencies. There was also uncertainty concerning the interpretation of existing federal laws which may apply in such cases.

The primary purpose of the 1986 Co-ordinated Framework for Regulation of Biotechnology in the USA was to clarify the responsibilities of the FDA, EPA and USDA. Where overlapping jurisdictions arise, the framework determines that one agency be designated the lead agency with primary responsibility, while the other is to be considered the secondary agency. However, the framework does not elaborate on the role of the secondary agency, and this could lead to conflicting interpretations without avoiding duplication.

The Biotechnology Science Co-ordination Act of 1986, which amended the FIFRA and the Toxic Substances Control Act (TSCA), clarified the regulations for the release of genetically engineered bacteria, viruses, fungi, cells and tissue and their use in manufacturing in the USA. The Bill amends the FIFRA by defining pesticide as not including a genetically engineered organism and thus no permit may be issued for any such organism being tested for pesticide qualities under the FIFRA. To the TSCA, a section has been added which governs the regulation of the release of genetically engineered organisms into the environment and the use of such organisms in manufacturing. However, requests to undertake field trials of living organisms which are not gene deletion mutants, or which are not designated pathogenic, or which are not intergeneric will not be referred for higher review by the EPA Biotechnology Advisory Panel (see 6.1 and 6.3) because current thinking among the regulatory authorities in the USA is that the environmental release of such organisms does not constitute a potential hazard to which special attention must be paid.

Several governmental departments and agencies in the UK are involved in the development and enforcement of regulations governing the contained use and deliberate release of genetic engineered organisms. These include the MAFF, which is responsible for regulations and guidelines covering plant, animal and human health, and the DoE, both of which have a dual role as both sponsor and regulator of biotechnology, and the Planned Release Subcommittee of the ACGM. This subcommittee uses Advisory Notes relating to the release of genetically manipulated organisms which were drawn up by the ACGM under the chairmanship of Sir Robert Williams, then submitted for approval by the Health and Safety Commission (HSC) and issued by the

HSE in April 1986. The ACGM Advisory Notes put the initial onus on the experimenter to demonstrate that adequate safety tests have been, or will be, carried out in advance of release and that there are adequate plans to monitor the release and to abort the experiment if it proves necessary. Proposals for the release of recombinant organisms will be expected to provide data concerning procedures used for manipulation, the nature of the changes, stability of the novel organism as well as predictions of possible effects on the ecosystem.

Currently only a few specific types of release experiments are governed by binding legislation in the UK. These are the release of genetically engineered organisms that are not native, for which permission is required from the Secretary of State under the Wildlife and Countryside Act, and the release of microbial pesticides, which are covered by existing pesticide legislation. Apart from these specific cases, there is no legislation in the UK actually forbidding the release of genetically engineered organisms. The Advisory Notes of the ACGM are a voluntary code, and it is not compulsory either to notify the ACGM of planned releases or to heed their advice. Moreover, the Advisory Notes include a clause which permits companies to keep certain commercially sensitive information secret.

The ACGM has requested legislation that would require mandatory notification of all proposals to release genetically manipulated organisms be given to the ACGM and to other bodies, such as local environmental health officers. Following consultation with the ACGM, the HSC proposed in a consultative document published in September 1987 that the Health and Safety (Genetic Manipulation) Regulations (1978) be revised to mandate the notification of the contained use and planned release of genetically manipulated organisms and broaden the definition of genetic manipulation and release (HSC, 1987; see also 6.1). These changes may not satisfy critics of the regulations because the new definition of genetic manipulation does not include gene deletion mutants, and the definition of release does not appear to encompass the use of live viral vaccines, which means that it might not be mandatory to notify the

ACGM of applications to use or release deletion mutants or live viral vaccines. There is also concern that the HSW Act (1974), to which the Health and Safety (Genetic Manipulation) Regulations are supplementary, is inadequate to address environmental considerations. For example, the categories of risk laid out in the ACGM Advisory Notes which determine whether notification is to be given prospectively or retrospectively are based on the 'pathogenicity risk-assessment' scheme proposed by Sydney Brenner (see 3.5) which defines hazard in terms of risk to human health. A decision on the proposed changes in legislation is pending a report of the Royal Commission on Environmental Pollution, chaired by Sir Jack Lewis, due to be published in 1988.

Concern about undue delay whilst the ACGM assesses proposed release experiments on a case-by-case basis has aroused the anxiety of potential industrial applicants, worried about the competitive advantage to foreign organizations conducting their field research outside the UK. ICI, for example, is urging the ACGM to identify categories of genetically manipulated crops that will be exempted from the notification procedure. The Department of Trade and Industry (DTI) has initiated a 'Planned Release of Selected and Manipulated Organisms' initiative. The initiative is to be a programme of experiments to assess the risk of the deliberate release of genetically manipulated plants and micro-organisms, for which the Biotechnology Unit of the DTI is prepared to meet half the costs (Newmark 1987b). A major objective of the initiative is to hasten the evolution of a framework for deliberate release in order to rationalize the assessment procedure by the ACGM, and perhaps to make it depart from careful case-by-case scrutiny of individual projects and adopt a US-style approach, in which applications to release certain categories of genetically manipulated organisms are not subjected to scrutiny by the EPAs Biotechnology Advisory Panel because the nature of the genetic modification to which they have been subjected is not considered to pose a particular environmental threat (see also 6.3). This is in marked contrast to the view taken by a commission of scientists, industrialists and parliamentarians in West Germany which, fearful of the potential for undesirable consequences and less confident about the powers of predictive ecology, recommended (in vain) that a five-year moratorium be imposed on all releases involving genetically manipulated micro-organisms.

The European Commission is expected to propose legislation for a notification and endorsement system for the contained use and planned release of all types of organism. It is expected that the proposal will be that having received notification of a proposed contained use or planned release of a genetically manipulated organism, a Member State will complete a risk-assessment, according to certain minimum test requirements, and then forward the notification dossier to the Commission and other Member States. If none of the other Member States raise objections to the proposed use or release, then it will be able to go ahead. However, if there is a conflict of opinion, the Commission would appoint a panel of scientists to take a decision. However, this scheme is unlikely to find favour with some industrialists. Not only is there likely to be long delays in the notification procedure, but the Member Countries will also have to unify their risk-assessment procedures, which would be unpopular should the West German recommendation of a five-year moratorium be accepted.

The history of releases of recombinant organisms is a catalogue of misunderstandings, injunctions and regulatory violations. Regulatory confusions, commercial pressure and the lack of adequate definitions of what constitutes containment and environmental release increase the likelihood that unauthorized environmental releases will take place.

Frustrated by regulatory constraints and public protests, fearful of bad press from court injunctions, and disgruntled by what they perceive of as lack of government agency co-ordination, genetic engineering companies have begun to conduct their release experiments involving recombinant organisms in countries where the obstacles appear to be fewer due to laxer legislation and lower public awareness.

The extreme view is that the risks of the release of

recombinant organisms are virtually non-existent (Robertson 1986). There are those who claim that the extra burden to genetically engineered organisms carrying new genes will decrease their ability to compete and persist in the environment and that existing organisms will compete successfully with any engineered organisms that might be released (Brill 1985; Davis 1987). However, were genetically engineered organisms unable to compete or survive, they would be unlikely to be effective in their intended role in the environment.

The concept of homeostasis, invoked as a biological metaphor for a self-regulating system, is sometimes misconceived. Although ecosystems possess a certain degree of homeostasis, stability or resistance to change, and can often restore their function following a disturbance, they are in a permanent state of dynamic equilibrium. They are not perfectly self-regulating; the evolutionary history of the biosphere is characterized by dramatic fluctuations in the population size of different species, fluctuations which occur independently of human activities. Although full explanations are not available, there are some examples in which a small amount of genetic change seems to be the key factor (Halvorson *et al.* 1985).

Chance mutations occurring in single individuals have a low probability of exploiting ecosystems fully and becoming established. In commercial environmental release experiments, modified organisms designed to perpetuate in the environment will be released *en masse* and may be able to exploit the spectrum of niches available to them in ecosystems by virtue of their genetic modifications. Genetically engineered organisms have the potential to inflict deleterious and unanticipated effects on the structure and function of the ecosystems to which they are applied and to other ecosystems to which they may migrate.

There are those who maintain that the potential benefits to be derived from the environmental release of recombinant organisms outweigh the risks (Szybalski 1985). None the less, the insurance industry in the USA has been unwilling to provide insurance cover against the event of mishaps in field trials of recombinant microbes, and some countries have exercised laudable caution; Denmark has a total ban on the release of genetically engineered organisms, and a commission in West Germany proposed a temporary moratorium.

A large portion of any benefits derived from the use of this new technology to engineer organisms that are released *en masse* into the environment will undoubtedly accrue to the producers, and it is with the producers that the responsibility and liability for ecological catastrophe should also lie. Prior to the release of recombinant organisms there is a need for contingency plans in case of unforeseen undesirable consequences. In addition to which, there should be severe economic penalties and the withdrawal of permits where there has been violation of procedures laid down for the use of genetically engineered organisms.

Prior to the application for the release of any recombinant organism, a risk-assessment should be undertaken by the applicant taking into account the potential for survival, growth and reproduction of the organism at the location of release, the potential for transport to other locations, the potential for survival, growth and reproduction at such other locations, and the potential for the transfer of genetic materials from the organism to other organisms. There is, however, little information at present on the properties of an organism or an ecosystem which limits the success of the introduction of a foreign organism.

There is an analogy between the release of genetically engineered organisms into the environment and the introduction of exotic (non-native) species, whether accidentally or intentionally, into new environments, the ecological effects of which have been studied (Mooney and Drake 1986). Although many introduced exotic species do not survive in their new environment, the ones that do survive cause changes which range from the barely discernible, to obvious and extensive changes, including the extinction of desired varieties or overpopulation of undesirable ones. Ecologists often remain unable to explain, even after the event, why certain species deliberately introduced into foreign ecosystems became established or became pests (Mayer, 1987). These cases involved the release of non-genetically engineered organisms whose natural history can, in theory, be studied prior to their release in a foreign environment. The genetically engineered organisms for which applications for deliberate release are sought have either never existed before or have not been released into the environment in question *en masse* before. Our capacity to predict accurately the effects of the deliberate release of novel microgenetically engineered organisms is very limited and releases involving micro-organisms are regarded as the most difficult to assess. The lesson to be learnt from the results of accidental or deliberate release of animals, plants or microbes into novel environments is that present ecological modelling is inadequate to predict with accuracy the outcome of such releases. Until the science of predictive ecology is better developed, it is premature to exempt applications to release certain genetically engineered organisms from the review procedure.

There is a need for administrative programmes to analyse in depth the ecological consequences of experiments involving the release of recombinant organisms. More extensive technology assessment, utilizing appropriate techniques, is essential if a basis for effective standards of safety for the release of recombinant organisms is to be established. Indeed, neither adequate promotion nor prohibition of recombinant DNA technology are possible without more extensive technology assessment.

In EEC countries, policies of more systematic science and technology priority assessment are increasingly being introduced. In the UK, *ad hoc* commissioned consultancy reports, advisory committees and public enquiries assist governments and the general public to assess technological developments. Royal commissions are often appointed to investigate major issues, especially in cases where the public has been alarmed by the occurrence of some accident or disaster. The Royal Commission on Environmental Pollution, due to report on the environmental release of genetically engineered organisms in 1988, is the only permanent one. The USA is unique in having a Technology Assessment Act, which was passed in 1972. This Act established the Office of Technology Assessment (OTA) with the purpose of ensuring that technology is more socially acceptable and to inform legislators better about technical and scientific matters.

One legacy of the recombinant DNA debate of the 1970s is that scientists are unwilling to accept regulatory controls based upon hypothetical scenarios. At present, the science of ecological risk-assessment is immature. Ecologists need time to develop better models for the analysis of ecosystems and to devise better computer simulations to permit predictions of greater accuracy of the dynamics of ecosystems in nature, with their multitudinous interacting components (Abraham and Shaw 1981).

The scientific community and genetic engineering firms should exercise restraint in pursuing release experiments until such time as the science of ecology has matured sufficiently to provide the methodology adequate to the task of risk-assessment. In the interim, it would be responsible to have a worldwide moratorium of at least five years' duration.

NOTE

[1] The PAHO bovine rabies vacine experiment was terminated by the Argentinian government in September 1986. The Argentinian Ministry of Health alleged that farm hands who cared for the vaccinated cows and cows that were not vaccinated had antibodies to the mutant vaccinia indicating that they had been infected with the live vaccine.

8 Biowar

Microgenetic engineering permits the construction of microbes with previously unattainable combinations of characteristics. In this chapter we focus our attention on the ways in which the advent of this new technology has added a potentially catastrophic dimension to biological warfare, revolutionizing both defensive and offensive biological weapons research.

8.1 AGENTS OF BIOLOGICAL WARFARE

The agents of biological warfare are bacteria, viruses, fungi and rickettsia, a form of bacteria. Potential biological weapons agents are listed in Table 8.1.1, from which it is clear that there are agents to suit most 'requirements', their nefarious effects ranging from a high mortality rate to temporary incapacitation. The target organism may be people, livestock or crops.

Microgenetic engineering can be used to produce highly pathogenic organisms, for example, novel plant pathogens capable of destroying whole regions of cultivated crops, and novel strains of common infectious agents which render existing vaccines, the body's immune system and most antibiotics, useless. The application of this new technology to create new or better vaccines for use by the military is considered by some to be both offensive and defensive research. A country in possession of a vaccine against its own biological agents of warfare, with which it can innoculate its own troops and civilians, is more likely to launch an offensive, and the capacity to mass-produce vaccines against the biological weapons of an opposing army reduces the fear of retaliation. Some common forms of vaccine production are very close technically to the production of biological warfare agents and so offer easy opportunities for conversion (SIPRI 1973).

Research into biological agents of warfare between 1940 and 1969 was directed towards producing a strategic weapon. However, a study group of the World Health Organisation (WHO) reported in 1970 that the use of biological warfare on a strategic scale was probably impractical: there were too many risks of the attack backfiring (WHO 1970). For example, there is the danger that the agent of disease will mutate to an unpredictable new strain against which the aggressor may not be immunized. Another problem is that biological weapons may be persistent. For example, anthrax spores contaminate the environment for decades. The island of Gruinard off the north-west coast of Scotland remained contaminated for almost half a century after the anthrax experiment conducted there by British military research scientists from Porton Down in 1944 during the Second World War (Seagrave 1982). A programme to decontaminate the soil was undertaken in 1986 by the Ministry of Defence. The USA maintains that an explosion at a biological weapons laboratory in Sverdlovsk in the USSR in 1979 caused an outbreak of anthrax (S. Rose 1987). Recurrent attacks are a danger when using contagious viruses or bacteria which may generate a reservoir of disease in a host animal species, thereby creating a new source of infection and thus keeping an area constantly dangerous. In the 1960s the University of Utah conducted secret experiments for the US Army at Dugway Proving Ground, which involved large-scale open-field testing of some of the most infectious and toxic agents of biological warfare, including Rocky Mountain spotted fever, plague and Q-fever. The purpose of these field tests was to track the spread of these diseases over wide areas, and involved infecting thousands of wild animals using insects and aerosol chambers to disseminate the infectious agents (Piller 1985).

Table 8.1.1: *Pathogenic micro-organisms studied as potential biological warfare agents*

Disease/military target and type	Causative agent	Death rate in untreated cases of natural disease (%)	Likely mode of dissemination	Remarks
Anti-personnel agents				
Viruses				
influenza		0–1	aerosol	
psittacosis	*Chlamydia psittaci*	10–100	aerosol	
Russian spring-summer encephalitis	RSSE virus	0–30	aerosol or tick vectors	
yellow fever		4–100	aerosol or mosquito vectors	standardized as BW agent by US*
dengue fever		0–1	aerosol or mosquito vectors	once infected, mosquito bite can cause disease
chikungunya		0–1	aerosol or mosquito vectors	
Venezuelan equine encephalomyelitis	VEE virus	0–2	aerosol or mosquito vectors	standardized as BW agent by US*
Rift Valley fever	RVF virus	0–1	aerosol or mosquito vectors	
smallpox	*Variola*	0–30	aerosol	Japanese tests on prisoners-of-war**
haemorrhagic dengue		0–5	aerosol or mosquito vectors	Japanese tests on prisoners-of-war**

Rickettsiae				
epidemic typhus		10–40	aerosol	Japanese tests on prisoners-of-war**
Rocky Mountain spotted fever		10–30	aerosol or tick vectors	standardized as BW agent by US*
Q fever	*Coxiella burnetii*	1–4	aerosol	Japanese tests on prisoners-of-war**
tsutsugamushi (scrub typhus)	*Rickettsia tsutsugamushi*		mite vectors	
Bacteria				
plague	*Pasteurella pestis*	30–100	aerosol or flea vectors	allegedly used by Japanese in China
anthrax	*Bacillus anthracis*	95–100	aerosol	standardized as BW agent by US;* intensively studied during Second World War by US and UK; Japanese tests on prisoners-of-war**
glanders	*Actinobacillus mallei*	90–100	aerosol	3200 inhaled bacteria may cause disease; Japanese tests on prisoners-of-war**
meliodosis	*Pseudomonas pseudomallei*	95–100	aerosol	
cholera	*Vibrio comma*	10–80	water contamination	like salmonella and shigella, allegedly used by saboteurs in China and Manchuria during late 1930s; Japanese tests on prisoners-of-war**

Continued

Table 8.1.1: *(Continued)*

Disease/military target and type	Causative agent	Death rate in untreated cases of natural disease (%)	Likely mode of dissemination	Remarks
typhoid	*Salmonella typhosa*	4–20	aerosol or water or food contamination	100 ingested bacteria may cause disease; Japanese tests on prisoners-of-war**
dysentery	*Shigella* (species)	2–20	water or food contamination	oral infectious dose is about 5000 organisms in the case of *Sh. flexneri*; Japanese tests on prisoners-of-war**
tularaemia	*Francisella tularensis*	0–60	aerosol	standardized as BW agent by US* Japanese tests on prisoners-of-war**
brucellosis	*Brucella* (species)	2–5	aerosol	1300 inhaled bacteria may cause disease; *B. suis* standardized as BW agent by US; intensively studied during Second World War by US and UK; Japanese tests on prisoners-of-war**
gas gangrene	*Clostridium perfringens*			Japanese tests on prisoners-of-war**
Fungi coccidioidomycosis	*Coccidioides immitis*	0–50	aerosol	

Anti-plant agents

Viruses

tobacco mosaic			airborne transmission occurs naturally
sugar beet curly-top			vector-transmitted naturally (leafhoppers)
corn stunt			
boja blanca (rice)			
Fiji disease (sugar cane)			
potato yellow dwarf			vector-transmitted naturally (leafhoppers)

Bacteria

rice blight	*Xanthomonas oryzae*		
corn blight	*Pseudomonas alboprecipitans*		
sugar cane wilt (gumming disease)	*Xanthomonas vasculorum*		natural windborne transmission observed

Fungi

late blight of potato	*Phytophthora infestans*	aerosol or dust	responsible for the Irish potato famine of 1845–69
coffee rust	*Hemileia vastatrix*	aerosol or dust	responsible for the elimination of coffee from Ceylon in 1880s
maize rust	*Puccinia polysora*	aerosol or dust	transatlantic airborne transmission has been observed

Continued

Table 8.1.1: *(Continued)*

Disease/military target and type	Causative agent	Death rate in untreated cases of natural disease (%)	Likely mode of dissemination	Remarks
powdery mildew of cereals	*Erysiphe graminis*		aerosol or dust	
black stem rust of cereals	*Puccinia graminis*	3–90	aerosol or dust	*P. graminis tritici* standardized as BW agent in US*
rice brown-spot disease	*Helminthosporium oryzae*		aerosol or dust	
rice blast	*Pyricularia oryzae*	50–90	aerosol or dust	standardized as BW agent in US*
stripe rust of cereals	*Puccinia glumarum*		aerosol or dust	
Anti-animal agents **Viruses**				
foot-and-mouth disease (cattle)	FMD virus	3–85	aerosol or water or food contamination	
rinderpest, or cattle plague		15–95	aerosol or water or food contamination	intensively studied during Second World War
vesicular stomatitis (cattle)		15–95	aerosol or water or food contamination	

Newcastle disease (poultry)		10–100	aerosol or water or food contamination	intensively studied during Second World War
fowl plague		90–100	aerosol or water or food contamination	a rare infection of chickens caused by a strain of influenza virus
African swine fever		95–100	aerosol or water or food contamination	
hog cholera		80–90	aerosol or water or food contamination	
Rickettsiae				
heart-water, or Veldt sickness (sheep and goats)	*Rickettsia ruminantium*	50–60	aerosol or tick vectors	
Fungi				
aspergillosis (poultry)	*Aspergillus fumigatus*	50–90	dust or food contamination	
lumpy-jaw (cattle)	*Actinomyces bovis*	50–90	food contamination	

Notes: * Biological agents which have been standardized or manufactured for inclusion in certain national chemical and biological weapons arsenals in the past.
** John W. Powell, 'Japan's Germ Warfare: The US Cover-up of a War Crime', *Bulletin of Concerned Asian Scholars* 12(4), 1980.

Source: S. Murphy *et al.*, *No Fire, No Thunder* (London: Pluto, 1984). With the permission of the publisher.

The present emphasis is on the production of undercover tactical agents against selective military targets with the aim of slowly debilitating the opposition (S. Wright 1985). Any country which is reliant on monoculture agriculture for cash-crop exports is particularly vulnerable to biological sabotage. The close resemblance between naturally and unnaturally occurring epidemics may divert suspicion away from those caused by clandestine acts of terrorism using biological weapons.

Sometimes serious disease is caused not by a micro-organism itself but by one of its harmful products known as a toxin. Diphtheria and tetanus are examples of such diseases caused by micro-organisms multiplying locally and producing potent toxins with life-threatening effects on the heart and nervous system. Toxins have also been employed as weapons. Botulin and ricin (a toxic protein synthesized by the castor bean) were researched during the Second World War. More subtle than biological agents, toxins can be used to kill individuals rather than large numbers of people.

A previously unimmunized individual can be protected against a toxin by intramuscular injection of an antitoxin—a serum containing high levels of antibody to the toxin. Alternatively an individual may be vaccinated with a toxoid, which is made from purified toxin which has been inactivated with formalin to create a substance which is non-pathogenic but antigenic. The new techniques of genetic engineering enable antitoxins and toxoids to be synthesized rather than extracted. For example, hybridoma technology can be applied to create large quantities of specific antibodies for use as antitoxins and recombinant DNA technology can be applied in the synthesis of toxoid vaccines.

8.2   PROTECTION AGAINST BIOLOGICAL WEAPONS

There are three main forms of defence against the strategic use of biological weapons in a military offensive. One method is to enhance the resistance of the target population, be it people, livestock or crops, to infections by the biological agent employed or its harmful effects. Livestock and people can be protected from biological weapons by the use of specific antibiotics, antitoxins and vaccines. However, large-scale civilian vaccination for rare or exotic diseases would create public anxiety and would also alert adversaries to the defensive (and, possibly, offensive) biological weapons strategies being developed. Theoretically, an entire civilian population could be covertly innoculated against biological weapons agents by spraying a vaccine in aerosol form over wide areas. In aerosol immunization, the vaccine against such a potential biological warfare agent as anthrax is inhaled rather than swallowed or injected. An inherent risk in any vaccination programme which uses a live vaccine is that the vaccine will mutate to a harmful agent of disease. Aerosol immunization is more dangerous than standard methods since the vaccine is projected into the atmosphere.

Since the Second World War the US Army has been studying the effectiveness of aerosol immunization. In the 1950s the Army secretly disseminated 'simulants' in a number of civilian areas, including the San Francisco Bay Area and in New York City subways, to test the feasibility of this techniques for use in biological warfare. Revelations of the experiments were not made until the late 1970s (Harris and Paxman 1982).

In order to use vaccines, antitoxins and antibiotics as a defensive policy, specific protective regimes would have to be prepared against each agent which might be used by the opposing forces in an attack. As the number of potential agents which may be employed is, through the application of microgenetic manipulation, theoretically infinite, it is impracticable to take precautionary measures against all eventualities, even if it were possible to predict what they would be.

Two others types of defensive measures do not require such precise advance knowledge of the offensive agent to be employed. They are the use of protective clothing and early-warning systems capable of detecting biological weapons attacks, measures which would be effective against a wide range of potential offensive biological agents. The reason for their broad applicability is that the factors relevant to the efficacy or otherwise of these defensive measures are not dependent on the *biological* properties of the agent employed, but depend rather on the *physical* characteristics of their method of dispersal. Most biological agents are dispersed in the form of an aerosol, the important characteristics of which, for defensive work, are particle size and surface tension which can be matched with harmless simulants.

Early-warning systems and protective clothing can be developed and tested through simulation experiments using aerosols of non-pathogenic or avirulent microbes which replicate key molecular properties shared by numerous bacteria or viruses. On the basis of the 'worst case' assumption that contamination of the environment after an attack will be persistent, a decontamination strategy for normalization can be formulated without prior knowledge of the biological characteristics of the offensive agent.

Horror at the use of chemical warfare in Indo-China helped to fuel a widespread campaign in the 1960s to outlaw such weapons. A 1969 United Nations (UN) resolution affirmed that the Geneva Protocol prohibited the deployment of chemical agents of warfare. The Geneva Protocol of 1925 is a comprehensive treaty formulated under the League of Nations, and signed by nearly forty countries. It prohibits the use in war of asphyxiating, poison or other gases, and of bacteriological methods of warfare. However, it only bans *first* use, so a country may retaliate in kind if attacked, and it does not ban *research* into development and *stockpiling* of chemical and bacterial weapons.

The pressure of public opinion resulted in the reinforcing of the Geneva Protocol with the Biological and Toxin Weapons Convention of 1972. Under the 1972 Convention, signatory nations pledge never to produce, under any circumstances, microbial or other biological agents, or toxins, of types and in quantities that have no justification for prophylactic, protective or other peaceful purposes, whatever their origin or method of production. The ban includes research to produce the necessary agents and the development and the stockpiling of these weapons. It is the only disarmament agreement in which the parties are required to forego possession of, as well as use of, weapons. Signatories also pledge never to acquire or attain equipment or means of delivery designed to use such agents or toxins for hostile purposes or in armed conflict. The UK, the USA and the USSR ratified the 1972 Biological Weapons Convention in 1975.

8.3 MILITARY RESEARCH

Since the late 1970s there has been renewed military interest in biological research, particularly recombinant DNA technology (Pettinan 1980; Robinson 1982).

In the UK, individual researchers have sometimes accepted support grants from the British MoD, NATO and the US Army, Navy and Airforce. As the British MoD does not publish lists of such contracts, they may only be identified through press announcements. It has been suggested that more than sixty of the contracts between colleges and universities initiated by the Chemical Research Establishment at Porton Down are for research into areas relevant to chemical and biological weapons (Rose 1987).

The US 'intelligence community' has a modest budget to enable it to monitor developments in biological arsenals and their possible use which might make the US and the NATO forces vulnerable to an attack. The Pentagon alleges that the USSR has an active research and development programme to investigate and evaluate the utility of biological weapons. There are believed to be at least seven biological warfare centres in the USSR that are under strict military control (Joyce 1984a). The US Department of Defense (DoD) and the Arms Control and Disarmament Agency allege that the USSR is manufacturing biological weapons in violation of the 1972 Biological Weapons Convention and that it is using genetic engineering for this purpose (Asinof 1984).

In 1984 John Birkner, an environmental biologist with the US Defense Intelligence Agency, advised scientists and companies in biotechnology that their work is considered sensitive among those who seek to keep US technology secret from the USSR. Henry Mitman, a senior official with the US Commerce Department's exports division has recounted several high-level meetings held in Washington DC in the early 1980s where government officials laid plans for controlling biological technology and data whose export could threaten national security. The Commerce Department formed a technical advisory committee to advise the US government. This committee is composed of experts from companies engaged in biotechnology, and whilst universities may present their views, academics do not sit on the panel.

The military establishment in the USA is planning to add various biotechnologies under the existing categories for biological and toxin materials of the 'Military Critical Technologies List', an unofficial and expanding compendium of sensitive technologies. Included in these additions are high containment at production scale, biohazard decontamination, high-capacity bioreactors and separation, solvent extraction and drying, micro-encapsulation, and very small spray nozzles (Joyce 1984a).

The 1972 Convention was given the force of law in the USA when the Senate ratified it in 1975. The DoD has provided the NIH-RAC with a list of DoD recombinant DNA research projects and a statement indicating that the DoD is complying with the 1972 Convention and the NIH Guidelines for Research Involving Recombinant DNA Molecules. In 1982 a resolution of the NIH-RAC stated that there was no reason to believe the DoD has not complied fully with the Convention and the NIH Guidelines (Milewski 1985).

In the late 1970s the USA annual budget for biological weapons defence research was about $15 million, but under the Reagan administration it more than tripled to at least $50 million, although some analysts place the figure at over $120 million (R. Smith, 1984a).

Since 1980 the US DoD has initiated over 100 projects involving recombinant DNA methods and monoclonal antibodies as aids towards the evolution of vaccines against biological warfare agents as part of the Biological Defense and Chemical Weapons Research Program. This is in addition to other DoD genetic engineering projects designed to assist in the defence against biological and chemical agents. Examples include the following for medical research: the development of vaccines against malaria, dengue fever, anthrax, rickettsia, Rift Valley fever virus, encephalitis, sleeping sickness and other diseases that troops might encounter; the cloning of the gene for the neurochemical acetylcholinesterase for possible use in therapy for nerve-agent poisoning; and the engineering of enzymes to decontaminate chemical warfare agents. The following projects for non-medical research have been initiated: the development of new polymers such as nylons, rubber and structural silk to make sturdier, lightweight materials; the microbial production of lubricants; the evaluation of recombinant marine micro-organisms as marine anti-fouling compounds; and the design of biosensors to monitor the presence of chemical warfare agents on the battle field (Joyce 1984a). The US DoD is also investigating early-warning systems for identifying biological weapons agents.

Although Fort Detrick in the USA remains one of the largest chemical and biological warfare research establishments in the world, in 1984 a Senate appropriations subcommittee on military construction agreed to reallocate substantial funds in the Pentagon budget for the construction of a controversial new Defense Department laboratory at Dugway Proving Ground in Utah, located 87 miles from Salt Lake City. The new facilities, which will cost about $10 million, are seen as part of an extensive renovation and expansion of chemical and biological weapons (CBW) research facilities for Dugway Proving Ground, the total cost of which is estimated at over 300 million dollars.

The new laboratory is the US Army's first laboratory devoted entirely to non-medical research (R. Smith 1984b). Designed to conduct tests involving highly infectious and lethal biological agents, the Utah laboratory incorporates P4, that is maximum, government physical containment safety standards. There are only four other P4 containment facilities in the USA, and they are operated by the National

Institutes of Health (NIH), the National Institute for Allergic and Infectious Diseases (NIAID), the Centers for Disease Control (CDC) and the Army Medical Research and Development Command (AMRDC).

The Utah laboratory is needed, according to the US Army's requisition for funds, 'to evaluate biological defense readiness' and test protective and detection apparatus 'by employing toxic micro-organisms and toxins requiring a level of containment and safety which would not be otherwise available in the Department of Defense' (Military Construction, Army, 1984).

The Army's non-medical P4 laboratory in Utah has been criticized by some prominent scientists because the Army attempted to get appropriations for the laboratory through an obscure legislative provision that effectively barred formal congressional voting, hearings and debates. The aura of surreptitiousness and secrecy surrounding the development of the laboratory has compounded concerns over the effects of developing disease-bearing organisms and mistrust of the motives of using genetic engineering for biological warfare research (Murray 1985).

The laboratory has also been criticized on the grounds that it may be used in contravention of the 1972 Convention. Normally, 'threat assessment' and 'equipment vulnerability' work is classified. According to some scientists, one way to allay suspicion concerning military research is to declare that all work at the laboratory will be unclassified. Sheldon Krimsky, author of *Genetic Alchemy* (1982), who formerly served on the NIH-RAC, suggests that at the very least, a list of every organism to be tested at the Utah P4 installation should be published openly (R. Smith 1984b). Indeed, it has been suggested that the information needed could be obtained by simulated pathogens with much less hazardous organisms. However, the DoD has stated that some of the laboratory's work will be kept secret, and this will be decided on a case-by-case basis.

Another means of allaying public concern may be to create an 'expert' panel capable of scrutinizing the laboratory's work. Royston Clowes, a molecular biologist at the University of Texas in Dallas, favours a legislative prohibition on the development of new toxins and suggests that any proposals to use recombinant DNA techniques at the laboratory should be subject to approval by the NIH-RAC. However, the NIH-RAC is not able to investigate the US Army's compliance with the 1972 Convention since it has neither the mandate nor the resources. Compliance with the NIH Guidelines on research involving recombinant DNA techniques is mandatory for the projects which it funds, but compliance with the Guidelines is voluntary for the military and private industry. The US DoD maintains that its policy is to comply fully with the NIH Guidelines. However, according to US Army records, the Illinois Institute of Technology conducted recombinant DNA research on nerve gas antidotes throughout 1982 without registering an IBC with the Office of Recombinant DNA Activities (ORDA) at the NIH (Piller 1985).

Many genetic engineers in the USA are becoming alarmed about the ambivalence of defensive and offensive research, arguing that the US DoD, which claims to be performing military biotechnology for defence purposes only, can readily convert new organisms into insiduously tailored pathogens. Richard Goldstein from the Harvard Medical School has pointed out that many of the current investigations are concerned with disease viruses that are extremely rare, for example, the Rift Valley fever virus, and that under the guise of defence the USA is preparing vaccines which are unlikely to be useful under any circumstances other than those of a biological warfare offensive (see Murphy *et al.* 1984). For example, in 1984 the NIH-RAC approved an experiment planned by a teaching hospital run by the DoD to clone the gene that codes for the toxin of Shigella viruses. These viruses are responsible for certain forms of dysentery, and the toxin is extremely potent. The NIH-RAC justify their decision by claiming that cloning the gene in question could pave the way for a vaccine for cholera as well as dysentery (Dixon 1984).

In 1982 Richard Goldstein and Richard Novick proposed that the NIH recombinant DNA Guidelines be amended to ban research on the production of biological weapons by molecular cloning. The response of the US government was that whilst it had no objection to the amendment in principle, it demanded a rewording that permitted genetic engineering for 'defensive' purposes (Murphy *et al.* 1984). Distrustful of the integrity of what they perceive as the military-industrial complex in the USA, Jonathan King, a molecular biologist at the Massachussetts Institute of Technology (MIT), and Susan Wright, an historian of science at the University of Michigan, are lobbying for a moratorium on any research into military uses of recombinant DNA technology. Included in their proposed ban would be studies of vaccines or antidotes designed to protect soldiers or civilians from biological attack (Joyce 1984a).

8.4   RESPONSIBLE SCIENCE

The present international law and specific conventions unambiguously outlaw the development, production, stock-piling, acquisition or retention of microbial or other biological agents or toxins, and their military uses. However, it is not at all certain that the existing conventions will restrain these developments. In the first place, not all countries are signatories to the specific conventions. Furthermore, without policing of military research laboratories by an independent international agency there can be no guarantee that covert production of biological weapons is not being undertaken.

The Conference on Disarmament has been seeking a treaty banning the production, storage and use of chemical weapons since 1968. However, it was the view of NATO officials at the 1986 Conference on Disarmament in Geneva that without what is called 'proper verification procedures' on production, storage or destruction of chemical weapons, any accord reached would be useless. If the escalation of biological weapons development is to be prevented, then it is imperative that signatories to the 1972 Convention form an international inspectorate composed of independent expert representatives from all the signatory nations to police international control of organized biological weapons manufacture, biological sabotage and terrorism. The UN should take on the responsibility of persuading non-signatory nations of the importance of joining with signatory nations in ensuring ratification of a total international ban on biological agents and toxins as set out in the Geneva Protocol and the 1972 Convention.

Unless the spirit of the international law and specific conventions relating to biological and toxic weapons is shared and positively and responsibly upheld by signatory nations, then there is always the possibility of reinterpretation and circumvention of its prohibitions in order that recombinant DNA technology may be used as a Trojan Horse with which to introduce offensive biologicals. It has been suggested that the techniques developed in molecular genetics, which enable the determination of the fine structure of the active sites of toxins, may inspire the design of weapons that might be produced entirely by chemical means, thereby enabling the military to trespass from the prohibited biological weapons area into chemical weapons, which have not yet yielded to years of disarmament efforts (Heden 1982; Harris and Paxman 1982).

Finally, biological agents can be produced under the 1972 Convention for 'prophylactic, protective or other peaceful purposes'. There is, however, little disagreement amongst scientists that the knowledge required to develop biological defensive measures to protect a population against biological attack is virtually indistinguishable from that needed to fabricate such agents for offensive use. The biosphere would be much less threatened if emphasis were placed on developing physical defensive measures, such as protective clothing and early-warning systems, which can be developed using harmless simulants and do not require the microgenetic engineering of dangerous pathogens in high-containment laboratories.

In the late 1960s increasing public concern with the integrity of the scientific enterprise gave rise to a 'radical science' movement. This movement included such groups as the Science for the People in the USA, the Campaign for Nuclear Disarmament (CND), the British Society for Social Responsibility in Science (BSSRS) and the Russell Foundation.

In the early 1980s the Russell Committee Against Chemical Weapons called on scientists and technologists not to participate in research associated with the development or production of chemical weapons and urged the British government to renounce such weapons and promote negotiations for their withdrawal from the European theatre of war (S. Rose 1981).

In 1983 the Committee for Responsible Genetics (CRG) was formed in Boston in the USA by a group of scientists and non-scientists. The aim of the CRG is to discuss and evaluate the social implications of biotechnology and genetics and to educate the public on these issues. In 1985 the CRG formed a subcommittee, the Committee on the Military Use of Biological Research (CMUBR), as part of a campaign to prevent the use of biology for weapons purposes. Other associations concerned about biological weapons developments are the Stockholm International of Peace Research Institute (SIPRI), Sweden, and the Nerve Centre, Berkeley, California.

Genetic engineering is Janus-faced, simultaneously regarding utopian and catastrophic scenarios. So far, most genetic engineering developments have been linked to the industrial rather than the military sector of the economy. Indeed, biologists have occupied an important role in moderating state-sponsored military developments by emphasizing the potential human suffering and damage to the environment from the use of technologically advanced forms of warfare. However, in the climate of terror produced by chemical warfare rearmament, this situation is rapidly changing. It has been argued that the scientist's code of conduct is forced to submit 'so easily to the imperative of the state' that scientists are 'alienated in their function from society' (Salomon 1973). More than ever before, the onus is on scientists to exercise conscious responsibility towards military biowar research and refuse to participate where it is in contravention of the *spirit* of the international laws and regulations controlling such research.

In the spirit of *glasnost*, it is crucial that further international discussion takes place immediately to establish the best means of controlling the application of genetic engineering for military purposes. In particular, it is imperative that the public becomes fully aware of this new biological dimension to the 'arms race', and that through public opinion, pressure be brought to bear upon governments to ban its use for military purposes in order to avert the ensuing catastrophic consequences.

Part III
Perfectability *Versus* Responsibility

9 Gene Therapy

Recombinant DNA technology empowers us to manipulate
the genetic material of living cells more precisely than has
previously been possible. The technology was pioneered
using microbial cells, but there are various ways in which
the genetic material of human cells can be manipulated
using recombinant DNA technology. For example, in gene
replacement or gene activation the cells manipulated could
be somatic cells or germ-cells and the genetic manipulation
may be undertaken for therapeutic or enhancement purposes. In this chapter we focus our attention on gene
replacement of human somatic cells for therapeutic purposes.

The pace of development of new medical technologies
far exceeds the resources available for their implementation.
In a world of scarce resources, difficult choices must be
made between available options which attempt to maximize
social welfare. However, once a new medical technology
has been developed, there is a tendency for its applications
to be explored and for its use to become imperative. Such
is the case with the medical use of genetic engineering
through which a range of techniques are being developed
for the microgenetic manipulation of human cells. In this
chapter we attempt to provide a technology assessment of
somatic cell gene replacement therapy.

9.1 HUMAN GENETIC MANIPULATION

Gene therapy is the correction of the deleterious effect(s) of a genetic disorder using the new techniques of genetic engineering. Gene enhancement, by comparison, uses the new techniques of genetic engineering to manipulate normal cells to enhance a trait which is perceived of as desirable. An example of gene enhancement is the insertion of additional growth hormone genes in order to increase the height of an individual who is not of restricted growth. Some scientists believe that they will be able to enhance such traits as intelligence or athleticism through the microgenetic manipulation of the genome. This is not technically feasible because proficiency in such traits is the result of a complex interaction of unknown genetic determinants and environment experiences.

There are two basic approaches to gene therapy. One approach is called gene activation therapy. The objective of gene activation therapy is to replace the activity of a defective gene by activating a dormant gene which has a similar function. For example, some genes specify products that are used only during embryonic or foetal stages of development; by the time the baby is born such genes have become 'shut down'. Human foetal globins, proteins that are subunits of the oxygen-carrying haemoglobin molecule, are specified by one such set of genes. Theoretically, various post-natal single-gene recessive disorders of the haemoglobin gene complex, such as sickle-cell anaemia and the various thalassaemias, could be treated if the dormant foetal globin genes could be reactivated.

The other main approach to gene therapy is gene replacement therapy. Gene replacement therapy is the insertion of additional genetic material into defective cells to replace a function which is effectively absent. Theoretically, in gene replacement therapy genetic material which is alien to the species may be introduced, for example genetic material from another species or a novel gene designed in the laboratory.

Eugenics is the science which deals with all the influences that improve and develop the inborn qualities of a race to the utmost advantage. Genetic manipulation of sex cells, fertilized eggs, or of cells that give rise to sex cells (the germ-line) has the potential to be eugenic. The criterion for success in eugenic gene therapy is that the genetic modification is of therapeutic value both to the patient and to his or her future offspring. Figure 9.1.1 depicts the steps required for successful eugenic gene replacement therapy on a newly fertilized egg.

The replacement gene would be required to integrate into the genome of the fertilized egg in the correct position in order that it be appropriately regulated. When the

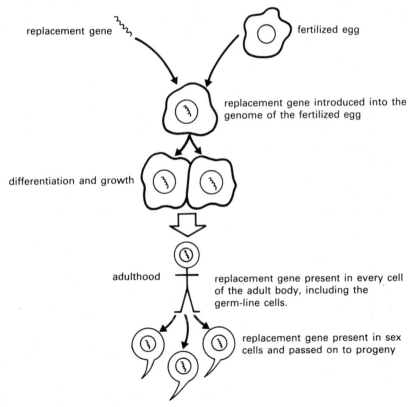

Figure 9.1.1 Eugenic gene therapy

fertilized egg replicates its DNA and divides, a copy of the replacement gene must be transmitted to each daughter cell, and this process must be repeated at every cell division in order that the replacement gene be present in every cell of the adult, where it will 'genetically fix' the disorder in the tissue(s) adversely affected by the genetic defect. Offspring of the successfully treated individual should inherit the replacement gene.

The first attempted eugenic genetic manipulation was the insertion into fertilized mouse eggs of DNA coding for rat growth hormone (J. Williams 1982). Rat growth hormone was detected in the resultant offspring in some cases, indicating that the rat gene had been integrated into the mouse genome and was being expressed. The expression of the rat gene was not regulated in a controlled way with the result that the genetically manipulated mice were extremely large and hence referred to as 'super mice'. The foreign gene was also present in the mouse germ-line cells, as evidenced by its expression in the progeny of the genetically engineered mice. A human DNA sequence has been used in a similar mammalian gene swapping experiment. A recombinant of rat growth hormone genes and human growth hormone genes was injected into the fertilized eggs of dwarf mice. The eggs were then implanted into female mice and some of the resultant progeny were 'cured' of dwarfism (Hammer *et al.* 1984).

Major technical advances are required before the insertion of genetic material into the germ-line of humans could be undertaken successfully. Micro-injection, the technique which has been used for eugenic manipulation in other mammals, is not presently acceptable for use in human patients because the procedure has an extremely high failure rate (Brinster *et al.* 1983). For instance, in 1987 a DNA molecule composed of the gene for a missing protein recombined with genes to regulate its expression was injected into fertilized mouse eggs in an attempt to correct the effects of a genetic defect known as 'shiverer mutation', which causes affected mice to die from uncontrollable shaking. From 320 mouse eggs, each injected with 200 copies of the recombinant DNA, only one mouse was born that was able to make the missing protein (I. Anderson 1987).

There is no control over where the injected DNA will integrate into the genome, and so its insertion site is random (Lacy *et al*. 1983). The success rate of eugenic gene therapy of mouse eggs using micro-injection is reduced by the fragility of fertilized eggs and by lethal mutations caused by insertion of the replacement gene into a critical part of the mouse genome. Before the dangers and efficacy of eugenic manipulation can be fully assessed, several generations of progeny should be studied.

The justification for mammalian gene swapping experiments is to enhance desirable properties in livestock and to undertake research that will be of use in attempting to cure genetic disorders in people. In the USA the environmental activist 'Jeremy Rifkin' and the scientific director of the Humane Society, Michael Fox, have proposed a prohibition on mammalian gene swapping experiments using human genes. They argue that the transfer of human genetic traits to other mammals transgresses the natural biological integrity of species. Rifkin also wants a ban on all attempts to engineer specific genetic traits into the germ-line of the human species. Although the NIH-RAC in the USA will not at present entertain proposals for human eugenic gene therapy, it unanimously rejected the proposed ban on mammalian gene transfer experiments (Joyce 1984b; Sun 1984).

Somatic cells are cells that are not involved in reproduction, and therefore genetic manipulation of somatic cells is not eugenic. The first recorded somatic cell gene therapy attempted in humans was when Martin Cline of the University of California in Los Angeles (UCLA) attempted in July 1980 to correct a defect in the bone marrow cells of two people suffering from beta-thalassaemia, a hereditary disease caused by a single base alteration in the DNA coding for the haemoglobin (oxygen-carrying) molecule of red blood cells. The attempt was unsuccessful and in June 1981 the NIH withdrew Cline's funding as a punishment for his premature experimentation without informed consent (see Wade 1981). (See also 5.2.)

Figure 9.1.2 outlines the stages for the application of somatic cell gene replacement therapy for the correction of a genetic defect which results in the absence or reduced level of a product of cells produced in the bone marrow. The replacement gene must be transferred to cells of the target tissue, in this case the bone marrow, where it must integrate into the genome in the correct orientation. The microgenetically engineered cells containing the replacement gene must be reintroduced into the patient's body, where the expression of the replacement gene must be regulated in the bone marrow so that the desired product is produced at the appropriate times and in the requisite quantity, and thus will alleviate the condition resulting from the genetic defect.

Preparations are underway in the USA for authorized clinical trials of human somatic cell gene replacement therapy (W. Anderson 1984). Points to consider in the design and submission of human somatic cell gene therapy

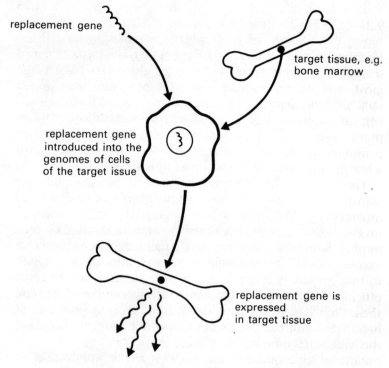

Figure 9.1.2 Somatic cell gene replacement therapy

protocols were issued by the NIH-RAC Human Gene Therapy Subcommittee in the USA in September 1986. A précis of each proposal is to be published in the *Federal Register* after which each proposal will be considered by the NIH-RAC on a case-by-case basis (ORDA 1986). The US congressional Office of Technology Assessment (OTA) report on the scientific and ethical issues inherent in the treatment of certain rare diseases by the repair or replacement of defective genes argue that there are no unique ethical obstacles to somatic cell therapy provided that considerations of safety and some reasonable expectation of efficacy are met. According to the OTA report, because cells that are used in reproduction are not involved, gene therapy in somatic cells is similar to other kinds of medical therapy and does not pose new kinds of risks (OTA 1984b).

9.2 COST-BENEFIT FRAMEWORK

Economics is the science of allocating given quantities of scarce resources among competing claims to obtain the most efficient or optimal use. As long as technology is not at a stage at which all wants can be satisfied, then ends will always be greater than the means available to achieve them. Policy decisions involve the allocation of scarce resources which have alternative uses, that is, they have what economists call 'opportunity costs'.

The choices made by medical practitioners and administrators in the allocation of scarce resources necessarily incorporate value-judgements. If a decision is made to proceed with a particular policy at a cost of £x, then implicitly the benefits are being valued at something in excess of £x. Economists have developed models to attempt to test socially efficient levels of output and for testing the efficient allocation of resources. Health economics applies these models to the allocation of scarce resources in healthcare and the use of health economics makes explicit the value-judgements underlying the choice of healthcare policy. In attempting to apply the principles of economics to policy decisions concerning alternative choices in healthcare, a common framework for assessment and choice must be devised to incorporate and evaluate the preference and judgements of those involved.

In assessing economic efficiency, the techniques of cost-benefit analysis have long been applied by economists to provide background information for decision-making. However, problems have arisen in the use of cost-benefit analysis because of the temptation to oversimplify, to forget the uncertainties and margins of error, and to emphasize quantifiable at the expense of non-quantifiable factors. Since the mid-1960s attempts have been made to improve such studies by including social costs in the calculations. Social benefits are private benefits plus external benefits and social costs are private costs plus external negative effects. In cost-benefit analysis there is a need to place a monetary value upon each social effect, positive for benefits and negative for detrimental effects, and to obtain a final net value by summing all the values and considering all parties concerned. In this way a comparison of the net social cost-benefits of competing uses of resources can be made which aids decision-makers. Social welfare is theoretically maximized when marginal social benefit equals marginal social cost.

Gene replacement therapy and gene activation therapy are both still at the developmental stage. Prior to the development and diffusion of somatic cell gene therapy, a technology assessment of alternative ways of managing serious genetic disorders should be undertaken. To aid decision-makers in developing a policy for the management of genetic disorders, the total costs and benefits of alternative courses of action should be estimated and compared.

The costs and benefits diagram in Figure 9.2.1 is presented as an initial framework for evaluating alternative therapies from a social welfare perspective. Interpreting benefits and costs in their broadest terms, benefits measures the potential of the individual to lead a normal life, and costs measures the cost to the patient, closely associated individuals, medical personnel involved and society at large. Benefits

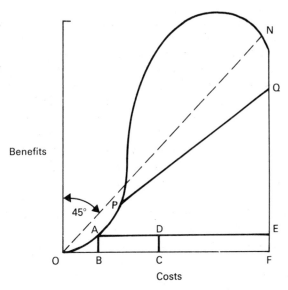

Figure 9.2.1 Cost-benefit framework for evaluating therapeutic medicine from a social welfare perspective

to the patient and society increase up the vertical axis, and costs increase along the horizontal axis.

A normal distribution curve under which each individual in a given population can be plotted with respect to his or her relative economic contribution (after Sewall Wright 1960) is superimposed on the cost-benefit diagram, 9.2.1. Area OAB represents those individuals in a normal population distribution who die *in utero* (in the womb) or very shortly after birth. Area ABCD represents those individuals who die before reaching maturity. Area CDEF represents those individuals with complete physical or mental incapacity throughout a lifetime of more or less normal length.

The 45-degree line ON shows where the cost-benefit ratios are unity at a given time. All ratios of benefits to costs below line ON would be sub-optimal from a social welfare point of view. Line PQ represents an arbitrary minimum threshold of cost-benefit ratios acceptable to society at a given time.

In order to use Figure 9.2.1 for evaluating alternative method of managing a genetic disorder, the social benefits and social costs of each alternative must be calculated and expressed as monetary values per (typical) patient. These values can then be plotted on the cost-benefit diagram. The relative positions of the points representing the social costs and benefits of each of the alternatives are then compared with each other and with line PQ. The implied decision-rule in choosing between the alternative methods of management of the genetic disorder from a social welfare perspective is that the preferred method be represented by a point somewhere above the line PQ. In the absence of alternatives above line PQ, the preferred method of management would be the one which minimises the social costs. In reality, this means deciding not to pursue the development of therapies which are deemed inadequately efficacious.

With gene therapy, as with any other novel therapy, in making the decision of whether or not to develop it and, if it is developed, whether or not to administer it in a particular case, the social costs and benefits of the treatment should be estimated. One approach is to estimate the total costs and benefits of alternative methods for the management of genetic disorders on a disorder-by-disorder basis. The rationale for this approach is that a number of relevant factors including: the severity of the untreated disorder; its incidence; the availability and cost of screening; and the availability, cost and efficacy of therapy; will vary greatly between different disorders.

To illustrate the cost-benefit approach we evaluate the alternative conventional methods of treatment of two genetic disorders, namely, phenylketonuria (PKU) and Lesch-Nyhan syndrome. Individuals with PKU are unable to metabolize a dietary constituent—an amino acid called phenylalanine. If untreated the condition causes a toxic accumulation of phenylalanine in the tissues—notably the brain—where it causes irreversible mental retardation and almost certainly results in institutionalization. The untreated sufferer can be expected to survive for about fifty years.

The cost-benefit position on Figure 9.2.1 of a typical untreated individual with PKU would be within the area CDEF. However, if the afflicted person is identified through neonatal screening for PKU and a low phenylalanine diet is initiated within the first month of life and continued for the duration of brain development—until between the ages of five and ten years—then there is a very high success rate in preventing mental retardation in PKU sufferers. Further dietary treatment is required only for females during pregnancy in order to prevent damage to the developing foetus (see Guthrie 1973; Hsia and Holtzman 1973). Table 9.2.1 summarizes the opportunity cost per individual institutionalized in the USA in 1972 as a result of brain damage caused by the failure to perform mass neonatal

Table 9.2.1: *Opportunity costs of screening for and treating phenylketo-nuria (PKU) in 1972 in the USA.*

	(US)$	(US)$
Institutional cost per untreated PKU individual (Over a period of fifty years at $20 per day)		365,000
Cost of screening per individual detected (14,000* tests × $1.25 per test)	17,000	
Cost of treatment per individual detected (Special diet for 5–10 years)	16,000	
Total cost of screening and treating per individual detected		33,000
Opportunity cost per individual of not screening for and treating PKU		332,000

Note:
*The incidence of PKU is 1:14,000 newborns

Source:
R. Guthrie, 'Mass Screening for Genetic Disease', in V.A.McKusick and R. Claiborne (eds), *Medical Genetics* (New York: HP Publishing, 1973).

screening (screening of the newly-born) and dietary therapy for PKU. As shown in Table 9.2.1, the financial cost-benefit ratio of neonatal screening for PKU followed by successful dietary therapy, as opposed to failure to screen and failure to treat, demonstrates that in economic terms, even including the expense of 13,999 negative tests to identify the one in 14,000 of the newborns with the condition, it is approximately ten times more costly not to treat than it is to screen and to treat. In addition to this saving, there is a beneficial contribution by the treated individual to social welfare through earnings, taxes and familial and societal contributions (Reilly 1977). The cost-benefit position on Figure 9.2.1 of a typical PKU sufferer after dietary treatment would fall in the area above line PQ, indicating that he or she are likely to make a significant contribution to society.

Our second example chosen to illustrate the cost-benefit approach to the evaluation of alternative therapies is that of chemotherapy for Lesch-Nyhan syndrome. Patients with Lesch-Nyhan syndrome are males because it is an X-linked recessive disorder. The biochemical basis of this disorder is a deficiency of the enzyme hypoxanthine-guanine phosphoribosyl transferase (HPRT), which characteristically results in the excretion of large quantities of uric acid. The central nervous system of boys with Lesch-Nyhan syndrome gradually deteriorates, and affected children are retarded, spastic, slow-growing and, for reasons that are not understood, the child is under compulsion to bit his lips and fingers so that his arms must be restrained to prevent self-mutilation. The condition is considered incurable, and few of those afflicted with it survive beyond their teens.

Without medical treatment a typical sufferer of Lesch-Nyhan syndrome would fall in the area ABCD of Figure 9.2.1. Therapy with allopurinol, which blocks the synthesis of uric acid, can reduce the serum levels of that substance, thereby minimizing kidney damage and extending the life of the child, but allopurinol does not alleviate the neurologic symptoms, which appear to be due to the anomalous production or accumulation of some compound other than uric acid that results from the same enzyme defect (Dancis

1973). The post-chemotherapeutic net social cost-benefit position of a typical sufferer of Lesch-Nyhan syndrome would fall into the area ABFE on Figure 9.2.1 which is below line PQ, which depicts our arbitrary socially acceptable minimum cost-benefit threshold. The balance between the social benefits of allopurinol chemotherapy for Lesch-Nyhan syndrome and the costs to the patient and the family and society in extended care and suffering make the decision to administer allopurinol therapy highly questionable from a social welfare perspective.

In assessing the social costs and benefits of developing somatic cell gene replacement therapy for the treatment of genetic disorders, the following questions should be addressed: What is the size and nature of the burden of genetic disorders? What are the costs of researching, developing and implementing somatic cell gene replacement therapy? What prospects of reducing this burden does this form of therapy offer? Are there alternative approaches to the relief of the burden of genetic disorders? What are the comparative costs and benefits of alternative approaches?

9.3 TECHNICAL MERITS

The social benefits of gene therapy arise out of its potential to reduce the burden of genetic disorders. In order to assess the potential social benefits of gene therapy it is necessary to assess the size and nature of the burden of genetic disorders and to assess the likely technical merit of gene therapy to alleviate it.

Genetic and environmental factors are involved in all disorders, although the relative importance of each varies from disorder to disorder and from case to case (McKusick 1969). Each disorder may be conceptualized as on a genetic (endogenous) and environmental (exogenous) spectrum with regard to the relative significance of genetic and environmental factors, as depicted in Figure 9.3.1. The medical profession classifies disorders into three groups with regard to the role of genetic factors: those caused by a single mutant gene; those caused through chromosomal

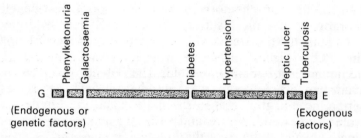

Figure 9.3.1 Continuum of disease indicating relative importance of genetic and environmental factors

Source: V.A. McKusick, *Human Genetics* (New Jersey: Prentice Hall, 1969), p. 182.

abnormality; and those which are multifactorial—that is, both genetic and environmental factors are multiple and interact in a complex manner.

Roughly one pregnancy in thirty will produce either a marked congenital malformation or a serious developmental abnormality which appears early in life. Common abnormalities include hare-lip, spina bifida, congenital heart disease and mental deficiency (Clarke 1974). The approximate distribution of those live births with some form of congenital abnormality among the three categories outlined above is summarized in Table 9.3.1.

Multifactorial conditions account for approximately 64 per cent of congenital abnormality and include congenital dislocation of the hip and spina bifida. Little is known of either the genetic or environmental factors involved in the causation of multifactorial conditions, and their correction

Table 9.3.1: *Causes of congenital abnormality*

Cause of congenital abnormality	Approximate distribution among congenitally abnormal live births (%)
Single mutant gene	12
Chromosomal aberration	24
Multifactorial	64

falls outside of the present technical capability of gene therapy.

Chromosomal aberration, as indicated in Table 9.3.1, is the second-largest cause of congenital abnormality and accounts for approximately 24 per cent of all cases. Down's syndrome, for example, occurs once in every 500 to 600 births, being more common among the children born to women late in their procreative life. It is the most frequent single definable entity causing severe mental deficiency. Down's syndrome is among the vast majority of viable chromosomal disorders which are caused by the presence of too much genetic material. Current methods of microgenetic manipulation do not permit the selective removal or 'switching off' of genetic material already present in a living cell, although this possibility has been brought closer by the introduction of laser technology into microgenetic engineering which permits the selective deletion of specific pieces of DNA (see Cross 1986).

Table 9.3.1 notes that approximately 12 per cent of congenital abnormality is the result of single-gene disorders. Although single-gene disorders are numerous in kind—over 1,600 have been identified—they are individually rare; over half a million children afflicted with single-gene disorders are born each year throughout the world (Watson *et al.* 1983).

Single-gene disorders are generally described as dominant or recessive. About 1,000 dominant disorders have been described. A dominant disorder is deleterious to an individual who inherits the mutant gene from just one parent. In a dominant disorder the mutant gene causes the synthesis of a substance which is deleterious to the organism, the presence of which overrides the action of the corresponding normal gene inherited from the other parent. Although referred to as the 'replacement' gene, the genetic material which will be transferred into the patient's cells in gene replacement therapy will be a supplement to the patient's genome; the mutant gene will remain *in situ* in addition to the inserted foreign gene (W. Anderson 1985). This prohibits its application in the treatment of dominant disorders in which the defective gene causes the synthesis of a deleterious substance. The potential of gene therapy is therefore restricted to the correction of single-gene disorders in which the product of the mutant gene is effectively absent. Such gene mutations are recessive.

The English physician, Archibald Garrod, observing earlier this century that several human hereditary traits, now known as recessive single-gene disorders, are characterized by the failure of known chemical reactions to take place, termed such conditions 'inborn errors of metabolism'. In one of the most significant hypotheses in human genetics Garrod suggested that such inborn errors were due to genetic blocks in chains of metabolic reactions (Garrod 1963). A recessive gene mutation results in the effective absence of a gene product which takes part in a metabolic pathway through low level of gene expression or through the production of an inactive form of the gene product. As a result of the metabolic lesion, the patient's body may accumulate one or more toxic chemicals causing severe mental and physical disabilities, or an essential bodily constituent, such as haemoglobin, growth hormone or blood clotting factor, may be absent or present only at reduced levels.

The gene mutations responsible for recessive traits are usually only harmful to individuals who do not have a corresponding normal gene. An individual who has one copy of the defective gene for a recessive disorder and one copy of the corresponding normal gene is a 'carrier' and is usually healthy, and it is to parents who are both carriers of the same recessive disorder that children affected by recessive disorders are usually born. The effective absence of the gene product of the mutant gene in a carrier individual is compensated for by the action of the corresponding normal gene inherited from the other parent.

There are two types of recessive disorders: X-linked and autosomal. In X-linked recessive disorders the mutation is of a gene on the X sex chromosome, and in autosomal recessive disorders the mutation is of a gene on an autosome (non-sex chromosome). Males inherit just one X chromosome and have no compensating copy and therefore X-linked recessive conditions are expressed almost exclusively in males. Females have two X chromosomes and will express an X-linked recessive disorder only if both parents pass on the trait. More than 100 X-linked recessive conditions are known, including haemophilia, colour blindness, Duchenne muscular dystrophy and Lesch-Nyhan syndrome. Classic haemophilia affects one in 10,000 males, for whom serious haemorrhage frequently results from relatively minor injuries or even spontaneously, owing to the possession of an abnormal blood-clotting protein. It results in pain, profound anaemia and orthopaedic problems caused by bleeding into the joints. Duchenne muscular dystrophy is a form of progressive wastage of the muscles which affects one in 3,000 male births. Apparently normal at birth, unusual clumsiness due to muscle lapse is apparent by 3 to 4 years of age, leading to wheelchair confinement by the age of 10, and death by about the age of 20.

Autosomal recessive conditions (in which the gene mutation is on an autosomal chromosome) occur in both sexes with equal frequency and are the most common type of single-gene disorder. Approximately 500 autosomal recessive conditions have been identified, including phenylketonuria (PKU), cystic fibrosis, albinism, sickle-cell anaemia and Tay-Sachs disease. Typically, they afflict babies and young people, and many of them are associated with chronic and distressing mental or physical handicap, or with both. Individuals with such disorders often die before reaching sexual maturity (Watson *et al.* 1983). Thus autosomal recessive disorders place a considerable burden on healthcare and other social services, and they also cause distress and suffering for afflicted children and their families (Weatherall 1985).

Examples of the frequency of affected individuals and carriers of several autosomal recessive conditions are given in Table 9.3.2. As Table 9.3.2 indicates, carriers of a deleterious trait are more frequent than one might expect from the frequency of those individuals who are affected by the disease. Another point to note is that in some populations certain recessive disorders have an extraordinarily high frequency. For example, the incidence of Tay-Sachs disease is ten times higher among Ashkenazi Jews than among other groups.

Somatic cell gene replacement therapy is likely to be restricted in its application to the treatment of individuals

Table 9.3.2: *Approximate incidence of some common recessive single-gene disorders*

Disorder	Approximate incidence among live births	Approximate incidence of carriers
Albinism	1:20,000 general population	1:71.9 general population
Alkaptonuria	1:1,000,000 general population	1:502.5 general population
Amaurotic family idiocy	1:40,000 general population	1:100.5 general population
Beta-thalassaemia	1:400 some Mediterranean populations	
Cystic fibrosis	1:2,500 Caucasians	1:25 Caucasians
Gaucher's disease	{ 1:2,500 Ashkenazi Jews { 1:75,000 others	
Phenylketonuria	{ 1:5,000 Celtic Irish { 1:14,000 others	1:25 others
Sickle-cell anaemia	1:500 US blacks	1:10 US blacks
Tay-Sachs disease	{ 1:3,600 Ashkenazi Jews { 1:35,000 others	1:30 Ashkenazi Jews

Sources: C.A. Clarke *Human Genetics and Medicine* (London:Edward Arnold 1974); P.R. Reilly *Genetics, Law and Social Policy* (Cambridge, Mass: Harvard University Press, 1977); J.D. Watson, *et al. Recombinant DNA: A Short Course* (New York: Freeman, 1983).

afflicted with single-gene recessive disorders associated with the reduced or absent function of a mutant gene. Most patients with recessive single-gene disorders are born to carrier parents who only become aware of their carrier status after the birth of an affected child. The development of widespread prenatal or neonatal screening would need to be an integral aspect of many gene therapy programmes for them to be effective in treating the disorder before irreversible damage occurred.

Each human somatic cell contains the complete genetic complement which is characteristic of the species and which includes between 50,000 and 100,000 pairs of genes. However, there is a high specialization of cell types, and in each cell type a unique array of genes is being expressed at any given stage of development. In order that the appropriate normal gene might be inserted, the mutant gene responsible for the disorder must be identified. This restricts the application of gene replacement therapy to those single-gene disorders for which the genetic basis has been elucidated. Once identified, a copy of the normal gene is cloned to obtain it in quantity, after which it must be delivered and expressed specifically in cells of the tissue(s) deficient in the gene product, and not in cells of other tissue types.

One approach to the problem of tissue-specificity is to isolate the target tissue from the rest of the body and manipulate it *in vitro* (i.e. in glass vessels on the laboratory bench). This approach limits gene therapy to the treatment of disorders of the bone marrow, and possibly the skin. Bone marrow has been found to be the easiest tissue to manipulate outside the body, and presently it is the only tissue from which cells could be extracted, grown in culture to allow insertion of exogenous genes, and then successfully re-implanted into the patient from whom the tissue was taken.

Once inserted, the replacement gene must be regulated so that it is expressed as a protein at the appropriate time and place, and in the correct quantity. It was at first anticipated that the haemoglobin gene complex would be one of the first areas of the human genome in which genetic defects would be corrected successfully by gene replacement therapy. The most common single-gene disorders are in the haemoglobin gene complex and affect the structure and function of the oxygen-carrying haemoglobin molecule of red blood cells. Sickle-cell anaemia and the various thalassaemias, which are single-gene recessive disorders of the haemoglobin gene complex, cause severe anaemia resulting in the deaths worldwide of about 200,000 children per year. The bone marrow contains the cells which give rise to red blood cells. However, it is now appreciated that the regulatory system that controls the synthesis of haemoglobin is particularly complicated and is not expected to be understood in the near future. Thus haemoglobin abnormalities will not be treated successfully by gene replacement therapy for some time, if ever, and early attempts at gene therapy will be restricted to the correction of defects in genes with a simple 'always on' form of regulation.

For the treatment of disorders such as Tay-Sachs disease, which affect tissues which cannot be removed, treated and replaced, an *in vivo* tissue-specific delivery system is required, possibly one which could be administered intravenously. A typical individual with Tay-Sachs disease is apparently healthy at birth and develops normally for the first three to six months. However, thereafter progressive degeneration of the nervous system follows, characterized by regression of the patient to a state of inability to crawl or sit unaided, cessation to react to stimuli and the onset of paralysis and blindness by eighteen months. Death occurs on average at forty months, after two or three years of intensive nursing care. The deficient enzyme— hexosaminidase A (Hex A)— in Tay-Sachs disease is absent from the brain, the liver, the kidneys, the skin, leucocytes (white blood cells), blood serum and cultured skin fibroblasts (cells) of patients (Kabach and O'Brien 1973). Some viruses are tissue-specific and may be effective as tissue-specific cloning vectors in gene delivery; for example, it is postulated that the hepatitis virus is a potential vector for the delivery of replacement genes specifically to the liver (see Vines 1986).

A major obstacle in somatic cell gene replacement therapy is that although animal cells can be made to incorporate new replacement genes, the genes are not recognized by the cellular apparatus that decodes them to produce the corresponding product, such as insulin, haemoglobin and interferon. 'Promoters' and 'enhancers' are DNA control sequences that increase the rate at which genes in their locality are transcribed into messenger RNA (mRNA) in the first stage of gene expression. It is speculated that promoters and enhancers which are tissue-specific in their action may at some future date overcome both the problem of tissue-specificity and the problem of the control of the expression of the replacement gene.

In gene replacement therapy it is necessary to express eucaryotic genes in eucaryotic cells. One vector currently under development for gene insertion into human cells is a retroviral system which uses mammalian retroviruses. Retroviruses are RNA tumour viruses that enter cells which are dividing and may insert a DNA copy of their genes into a site in the genome of the host cell. However, whilst the retroviral method is regarded by many as a method with great promise, it is limited by its inability to correct a defect in non-dividing cells, for example in brain cells.

In sum, it must be said that gene replacement therapy is not likely to benefit the majority of individuals with genetic disorders. It is unlikely to have an impact on the suffering caused by multifactorial, chromosomal and dominant disorders, which involve large numbers of genes or which require the selective removal of defective genes which have a deleterious effect. The applicability of somatic cell gene replacement therapy is limited to the correction of genetic defects which result in the absence or deficiency of a gene product, of which the genetics and biochemistry are understood, and for which the replacement gene has been isolated and cloned. The potential efficacy of gene replacement therapy as a treatment for such disorders is further restricted by its technical limitations. Some of these technical limitations, such as the imprecision of the delivery system, and the control of expression of the replacement gene, may be overcome with time, duly extending the range of recessive single-gene disorders to which somatic cell gene replacement therapy could potentially bring relief.

9.3   SOCIAL COSTS

Some of the potential social costs of gene replacement therapy are due to the risks inherent in the techniques. Once the replacement gene is integrated in cells inside the patient, the cells and the insertion agent are irretrievable. Underlying the approach to the treatment of genetic disorders through the insertion of a replacement gene is the reductionist notion that a human body with a single-gene disorder is like a machine with a broken part—the solution to the problem is to mend the broken part, that is, to insert a normal copy of the gene. However, the human body is a living dynamic organism which does not function like a machine, and the inserted normal gene will be an addition to, rather than a substitute for, the defective gene which remains *in situ*. It is not known what the influence of this combination of defective and normal genes will be, but there is a risk that the addition of supplementary genetic material will alter the cell's regulatory pathways inadvertantly affecting cell division or other properties, perhaps transforming it into a cancerous cell-line (W. Anderson 1985).

The optimal delivery system would insert the replacement gene into the target cell type(s) and would also direct it to a predetermined chromosomal site. The nucleus of each human somatic cell contains 50,000 to 100,000 pairs of genes arrayed on twenty-three pairs of chromosomes. The imprecise insertion of foreign genetic material could precipitate a deleterious result. For example, it is speculated that using retroviral gene insertion vectors one gene replacement therapy patient in five may succumb to leucaemia as a result of 'insertional mutagenesis'–the disruption of the function of target host cell genes by the integration of the retroviral vector (see Vines 1986).

Normal control of expression of gene inserts in mammalian cells growing in culture or in experimental animals has rarely been achieved. Uncontrolled expression of the replacement gene, genes of the viral vector, or of genetic elements in the host cell genome as a result of its disruption through retroviral integration, is a potential source of undesirable 'side effects'. Genes in the host genome and genes in the viral vector may share a common environmental 'trigger' which results in their expression. This is believed to be the case for HTLV-I and HTLV-II, the human T-cell lymphotropic viruses (HTLVs). HTLV retroviruses attack cells of the immune system called T cells, causing their uncontrolled growth. The regions controlling the expression of HTLV retroviral genes contain DNA sequences that are similar to the sequences of the control regions of two human genes–the interleukin-2 gene and the gamma-interferon gene. It is suggested that the same factors (probably in this case exposure to the antigens of invading foreign agents) that initiate the expression of the gene for interleukin-2 when T cells of the immune system are activated also initiate the expression of latent HTLV genes within the cell, the action of whose products leads to uncontrolled growth of the T cells (Marx 1986).

A particular hazard of the use of viruses as vectors in gene replacement therapy is the risk of the spread of the vector carrying the gene insert to cells other than the target cells. Spread of the recombinant vector through the patient's cells could result in its harmful presence in other tissues, including the germ-line, where it could affect future offspring of the patient. Escape of the recombinant vector from the patient could lead to the infection of other persons with whom the patient comes into contact.

The initial cloning, cutting and splicing of the DNA which is to be expressed in gene replacement therapy is performed using bacterial systems. In order to facilitate these manipulations, novel vectors—called 'shuttle vectors'—have been constructed that contain sequences enabling their replication and expression in both bacterial and eucaryotic cells. A shuttle vector which has been constructed for use in human gene replacement therapy is a hybrid between a bacterial vector and a mammalian retrovirus. When placed into specially modified mouse cells, this shuttle vector can become a retrovirus that is capable of infecting a broad variety of mammalian cells, including those of humans. As a result of their ability to infect both procaryotic and eucaryotic cells, shuttle vectors increase the risk of the accidental transmission of genes from bacteria to higher organisms.

For use as vectors in gene therapy, retroviruses are made biologically safe by making them deficient in their capacity to replicate and spread to other cells, but yet still able to enter the target cell and insert the replacement gene into its genome. In 1984 Richard Mulligan and his colleagues at the Massachussetts Institute of Technology (MIT) and David Nathan of the Children's Hospital in Boston micro-genetically engineered what they considered to be a 'safe' novel retroviral system which fulfils these requirements by combining the properties of two deficient strains of virus (Williams *et al.* 1984). The hybrid vectors are designed to undergo only a single round of infection; that is, in theory, they should no longer be infectious after entering the first cell. Unfortunately, a disabled retroviral gene insertion system may degenerate into an infectious agent. For example, spontaneous mutation during the preparation of disabled retroviruses may generate an infectious virus (Waneck 1985). There is a risk that gene replacement vectors will become a cancer hazard through genetic recombination or as a result of the activation of oncogenes caused by their random insertion into large numbers of cells. Alternatively, the exchange of genetic information between a disabled retroviral vector and an endogenous 'healthy' retrovirus could produce an infectious recombinant virus with properties which are unpredictable and potentially harmful to the patient and other people.

Gene replacement therapy through retroviral gene insertion is not merely therapy mediated through the application of a DNA or RNA product, it is a novel therapeutic process which has the potential to cause extremely undesirable 'side-effects' with irreversible consequences to the patient and possibly to other individuals.

It is argued that one of the social costs of the success of modern therapeutic medicine is the increase in the 'genetic load' carried by human populations (Muller 1935, 1963). It is claimed that the success of modern medicine in the elimination of infectious disease as a selective agent leads to the survival of those who would otherwise have died as a result of hereditary weakness, thus leading to dysgenics. A similar argument is made against somatic cell gene therapy; if somatic cell gene therapy is successful in circumventing the 'genetic death' (failure to reproduce) that would have occurred if 'natural selection' had been allowed to take its course, then there will be an increase in the population of the frequency of gene mutations. It is feared that this increase will cause a reduction in the 'genetic fitness' of the population.

Whilst successful somatic cell gene therapy will incur a social cost as a result of prolonging and increasing the quality of life of those afflicted with serious genetic disorders, it is not accurate to say that it will reduce the genetic fitness of the population. This is because the biological definition of fitness is the ability to contribute genetic material to the next generation: fitness in the biological sense is an elastic property, determined by genes and by environment. Therapy need not diminish fitness in the biological sense; its effect is to create for human beings an ambience in which genes and genetic make-ups no longer gravely diminish the fitness of their possessors (Medawar and Medawar 1977). However, successful somatic cell gene therapy will increase the frequency of those deleterious genes responsible for heritable genetic disorders which are treatable because it will increase the probability that individuals afflicted with such genetic disorders will live and contribute genetic material to the next generation. Thus the social cost to future generations of successful somatic cell gene therapy is not a genetic cost but an economic cost: a policy of somatic cell gene therapy is one of translating a genetic burden into a financial one, which will increase as this policy is pursued.

If scarce economic resources are used to develop and implement gene therapy, then the opportunity costs incurred are the potential social benefits deriving from alternative approaches to the management of genetic disorders which are mutually excluded by pursuing gene therapy. The major alternative therapeutic approaches to the management of genetic disorders are listed in Table 9.4.1 which includes a brief assessment of the limitations of each approach, together with examples of disorders to which each approach is applicable. Approximately fifty genetic disorders are currently treatable through the application of conventional forms of therapy, although the treatment is not always successful or satisfactory.

A general limitation to the successful treatment of genetic disorders, which also applies to gene therapy, is the occurrence of irreversible damage to tissue(s) prior to therapy. A social cost of a policy of therapy for genetic disorders is the cost of a policy of screening for affected individuals in order that they can be identified for treatment. In cases where the successful treatment of genetic disorders requires life-long therapy, a policy of therapy for genetic disorders has long-term social cost implications. A long-term social cost of a policy of therapy for genetic disorders is that successful treatment, by permitting affected individuals to survive and reproduce, increases the population frequency of the gene in question, possibly resulting in an increase in the number of affected individuals, causing a rise in the load on medical facilities and personnel (Howell 1974).

The cost-benefit of developing and applying appropriate therapy for serious genetic disorders should be compared with the cost-benefit of their prevention through investing in the provision of genetic counselling facilities and in research into diagnostic techniques and reproductive technology. Given the choice, many couples would prefer therapeutic abortion of an embryo known to be affected by a serious genetic disorder rather than to take on the financial and social consequences of rearing a child thus afflicted. Genetic screening for serious and effectively untreatable genetic disorders could be used to identify couples and embryos at risk. A couple thus identified as at risk could extend the range of genetic material at their disposal for procreation by employing, where appropriate, artificial insemination by a donor, egg donation and *in vitro* fertilization, thereby reducing the probability of fertilizing an affected egg. Prenatal screening of embryos could be followed by the option of therapeutic abortion of those found to be afflicted with a serious and effectively untreatable genetic disorder.

Lesch-Nyhan syndrome, together with the severe immunodeficiency diseases adenosine deaminase (ADA) deficiency and purine-nucleotide phosphorylase (PNP) deficiency are target disorders for early experimentation with gene replacement therapy using a retroviral gene insertion system. Children with severe combined immune deficiency eventually succumb to infections with microorganisms which generally lead to their death in early childhood. Bone marrow transplantation, thymic transplant and foetal liver transplant, along with immunoglobulin therapy have been attempted but with little success. In addition to which, there are immunological incompatibility problems with transplants, and bone marrow transplantation is a painful procedure for the patient and the donor.

ADA, PNP and Lesch-Nyhan syndrome are each characterized by the virtual absence of a specific enzyme in cells of the bone marrow. In some patients with severe immune deficiency there seems to be an absence of the enzyme adenosine deaminase (ADA) or purine nucleotide phosphorylase (PNP); in Lesch-Nyhan syndrome the missing enzyme is HPRT. It is into the extracted bone marrow of these patients that the replacement gene for the missing enzyme will be inserted in gene replacement therapy. In the case of all three disorders the normal gene has been cloned and is available. Although it is not technically possible to regulate the synthesis of the replacement gene's product, it is postulated that in these patients the production of a small percentage of the normal enzyme level should

Table 9.4.1: *Approaches to the treatment of genetic disorders*

Method of treatment	General comments	Disorder
Surgery	May leave unsatisfactory conditions such as partial blindness and paralysis in disorders such as retinoblastoma, hereditary cancer of the eye of the newborn.	Spina bifida Cleft palate Hare-lip Club foot Pyloric stenosis Retinoblastoma Glycogen storage disease
Dietary	The most effective method in the control of genetic disorders. Dietary treatment circumvents the affected metabolic step and may prevent the onset of damage. However, early identification is required.	Phenylketonuria (PKU) Orotic aciduria Cystinuria Maple syrup urine disease Homocystinuria Argininosuccinic aciduria Galactosaemia
Chemotherapy	Deterioration of the body organs, for example, the brain, prior to treatment. Such damage cannot be reversed.	Wilson's disease Orotic aciduria Crigler-Najjar syndrome

Transplant	Lack of donor organs, for example, the liver in Wilson's disease, and the kidney in Fabry's disease. Also immunological incompatibility problems resulting in delay which may lead to irreversible damage.	Lesch-Nyhan syndrome Cystinuria Wilson's disease Fabry's disease Adenosine deaminase deficiency
Enzyme therapy	Endeavours to alleviate the problems caused by 'inborn errors of metabolism' through the use of enzymes to reduce substrate levels and product levels accumulating in abnormal quantities. Problems arise in getting an exogenous enzyme into the brain; the high cost of purifying enzymes; and immunological rejection, although molecular genetics may alleviate the latter two problems.	Gaucher's disease Haemophilia Orotic aciduria Crigler-Najjar syndrome
Behavioural	Avoidance of drugs and environmental stimuli known to 'trigger' the condition	Sickle-cell anaemia Erythrocyte G6PD
Gene therapy	Currently an experimental mode of therapy with prospects for certain single-gene recessive disorders for which the biochemistry and genetics of the complaint are understood, appropriate tissue can be targeted, the gene has an 'always on' form of regulation, and is expressed in many types of cells. An irreversible mode of therapy which is potentially hazardous if retroviruses are used for the gene insertion. On the right are disorders considered suitable for treatment.	Adenosine deaminase deficiency Lesch-Nyhan syndrome Purine-nucleotide phosphorylase deficiency

be beneficial, and mild overproduction should not be harmful. However, the site at which the replacement gene will insert into the patient's genome remains uncertain.

The cost-benefit position on Figure 9.2.1 of a typical Lesch-Nyhan syndrome patient following gene replacement therapy is extremely unlikely to be above the line PQ, which represents the socially acceptable cost-benefit threshold. This is because the tissue most severely affected by the lack of HPRT, the enzyme deficiency which characterizes the disorder, is in the brain, and it is improbable that HPRT produced in microgenetically manipulated bone-marrow cells would reach the brain. The transplantation of non-microgenetically manipulated 'healthy' HPRT-producing bone marrow cells from a donor to a Lesch-Nyhan syndrome child, for example, did not improve the condition of the child (see Vines 1986). Thus, in addition to the technical uncertainties and potential hazards of somatic cell gene replacement therapy, the benefits in terms of life enhancement to a sufferer with Lesch-Nyhan syndrome of the synthesis of HPRT in the bone marrow remains unclear.

9.5 CAVEATS

In the absence of effective therapy, the suffering caused by genetic disorders to those afflicted and their families is great. Using conventional therapies only a minority of genetic disorders can be treated successfully, whereas correction of the effect(s) of a genetic defect in the tissue(s) of affected individuals through gene replacement therapy appears to offer relief from a number of single-gene recessive disorders.

A utopian view of the future of somatic cell gene therapy is that it will usurp the role of existing therapies for genetic disorders and extend the boundaries of therapeutic medicine, especially to treat many more of the 500 recessive single-gene disorders. The US legislative authorities are paving the way for it to be developed and its implementation in the correction of single-gene recessive disorders is forecast for the 1990s. Some scientists believe that undue caution and conservatism of ethics committees may defer or prevent the introduction of innovative treatments. After all, it is argued, Edward Jenner used smallpox vaccine and Louis Pasteur used rabies vaccine without the safeguards we demand today. As a consequence of the risks they took, immunization policies were developed whereby many dangerous and lethal diseases were eradicated (Motulsky 1984). It is suggested by some (see, for example Anderson 1984), that a balance can be reached between restrictive conservatism and recklessness. One of the criticisms brought against Cline by his peers was that the results his trials of gene therapy on mice gave no grounds for confidence that the therapy would be efficacious in humans.

Successful somatic cell gene therapy will increase the frequency of those deleterious gene mutations responsible for genetic disorders which are treatable owing to the fact that it will increase the probability that individuals afflicted with such disorders will live and contribute genetic material to the next generation. Thus the social cost to future generations of successful somatic cell gene therapy is not a genetic cost but an economic cost: a policy of somatic cell gene therapy is one of translating a genetic burden into a financial one, which will increase if this policy is pursued.

Whilst society should make every reasonable endeavour to alleviate suffering, and for some genetic disorders it may be appropriate to research and develop gene replacement therapy, given limited economic resources it is only right that efficient allocation of these resources for the social welfare be applied. When therapy is successfully applied for a genetic disorder there is a long-term social cost because the genetic load is increased. Moreover, gene replacement therapy through retroviral gene insertion has the potential to cause undesirable 'side-effects' with irreversible adverse consequences to the patient and possibly to other individuals.

Once developed, the use of medical technology tends to become an imperative, however expensive, ineffective and potentially dangerous it may prove to be. From a social welfare perspective, it is often safer and more economical to apply medical resources to preventive medicine rather than to attempt treatment or cure. A more socially cost-effective approach to the problem of the burden of genetic disorders is to reduce its size through preventive measures by investing in the provision of genetic couselling facilities and in research into diagnostic techniques and reproductive technology. Given limited health resources, the opportunity cost of pursuing a policy of gene therapy includes the social benefits foregone which would otherwise have been derived from the provision of adequate resources to support a policy of parental responsibility in preventive medicine. In furthering the development of gene therapy, the desire for perfectability must be tempered with responsibility.

10 Genetic Screening

Genetic screening encompasses a range of techniques used to diagnose phenotypic traits which have or are believed to have a genetic basis. Genetic screening is largely used to detect such traits before they become evident, but can also be used to verify the diagnosis of conditions after they have become apparent.

Techniques used in genetic screening include cytogenetic analysis, which is the examination of the genetic material at the chromosomal level for observable chromosomal abnormalities that are associated with particular genetic traits. Monoclonal antibodies, bacterial assays and biochemical tests are also used in genetic screening to detect cellular or biochemical abnormalities believed to be associated with abnormalities of the genetic material.

The techniques of microgenetic engineering, including restriction enzyme analysis, molecular hybridization, rapid DNA sequencing and cloning of DNA, have greatly increased the capability to examine, compare and analyse the genetic contents of cells. Microgenetic engineering has added a new dimension to genetic screening because it permits traits of the phenotype to be diagnosed through the analysis of variations in DNA. The genome is like a genetic blueprint and the only two individuals that have exactly the same genome are identical twins. Although we are composed of millions of cells, almost every single cell of our body contains a complete copy of our genome which means that genetic screening of our DNA can be performed using virtually any cells taken from our body. The DNA contained in the white blood cells taken from a 5ml blood sample, for example, is sufficient material for the application of any existing genetic screening test of DNA.

Microgenetic engineering, hybridoma technology and biochemical analysis are powerful new tools with which to screen for genetic variations and mutations in embryos *in utero*, babies at birth and adult men and women. Genetic screening is gaining increasing acceptance among public health authorities in the management of genetic disorders. Broadly speaking, the two alternative courses of action in the management of serious genetic disorders are treatment and prevention, and genetic screening plays a part in both. Genetic screening enables the early diagnosis of certain hereditary disorders which can only be treated successfully through early intervention before their effects are clinically manifested. The management of genetic disorders through prevention involves the genetic screening of prospective parents together with the options of germinal choice using adoption or the new reproductive technologies, or prenatal screening together with therapeutic abortion of affected foetuses.

As the spectrum of traits for which screening tests are available expands, genetic screening is also increasingly being used by non-medical agents in society, such as employers, immigration authorities, the law courts, the police and insurance companies. However, the implementation of genetic screening raises significant social, ethical and economic issues which warrant consideration prior to its further diffusion. In this chapter the methodology of genetic screening, its applications and the ensuing social ramifications of its applications are explored.

10.1   TEST RELIABILITY

The first requirement of a good genetic screening test is that it should be of proven reliability. A false negative test result is the failure to identify an individual with the condition screened for. A false positive test result is the identification of an individual who does not, in fact, have the condition. If a 99 per cent accurate screening test with a bias towards false positive results were applied without error to 1,000 individuals of a statistically normally distributed population, to screen for a condition which occurs in one birth in a 1,000, then although the one individual with the condition should be identified, it is possible that so would ten others be falsely identified.

The basis of unreliable test results from genetic screening may be due to inaccuracies in performing the tests, especially when the test methodology is complex. For example, the 'blue diaper test', a biochemical test for phenylketonuria (PKU) which was used in England from the mid-1950s until 1964, was unsatisfactory in several ways. Individuals with PKU are unable to metabolize a dietary constituent, an amino acid called phenylalanine. If untreated, the condition causes a toxic accumulation of phenylalanine in the tissues, notably the brain, where it causes irreversible mental retardation. The untreated sufferer can be expected to survive for about fifty years. A low phenylalanine diet initiated within the first month of life and continued for the duration of brain development—until between the ages of five and ten years—is usually successful in preventing mental retardation in PKU sufferers; further dietary treatment is required only for females during pregnancy in order to prevent damage to the developing foetus (Guthrie 1973; Hsia and Holtzman 1973). Unfortunately, it was not possible to perform the 'blue diaper test' before the baby was one month old, by which time the baby may have already suffered brain damage. Earlier testing produced as many as 45 per cent false negative results, which resulted in failure to treat the condition and led to irreversible brain damage and institutionalization of such children.

In addition to cleaving DNA in recombinant DNA technology, restriction enzymes provide a new and extremely valuable method for analysing genetic diversity. There are over 300 restriction enzymes available, and each cuts DNA at a specific site, the restriction site, which is composed of a sequence of four to ten bases. Treatment with a particular restriction enzyme will cleave a particular gene into a characteristic set of 'restriction fragments' of different lengths. Scattered throughout the human genome there are harmless mutations which may produce new restriction sites or remove pre-existing sites. Restriction site mutations of this sort would be reflected in a change in the length of one or more restriction fragments from that seen typically. The variability in the lengths of restriction fragments is called restriction fragment length polymorphism (RFLP) and is used in the analysis of variation in DNA for genetic screening.

Another method of genetic screening of DNA exploits the base-pairing that occurs between single strands of DNA that are complementary in base sequence. Short radioactively labelled single-stranded pieces of DNA, called DNA 'probes', will seek out, bind to and thereby label single-stranded complementary DNA sequences in the DNA under analysis. Researchers hope to identify and synthesize an array of probes to correspond to gene mutations associated with or responsible for a range of inherited conditions (see Weatherall 1985).

The problem with the use of gene probes is that mismatch can occur which results in the pairing of a gene probe with a single-stranded DNA molecule to which it is almost, but not quite, complementary. This mismatch can give rise to a genetic mis-screening. It has been found that short gene probes (called oligonucleotide probes) of about nineteen nucleotide bases are more reliable at detecting single base changes than probes which are longer. The ability to distinguish between single-stranded DNA molecules that differ by just one base is important because some genetic disorders are the result of a mutation in a single base.

Ideally, a probe used for a genetic screening test for an inherited disorder would be complementary to the gene mutation responsible for the disorder. However, in some cases the probe corresponds to a 'linked genetic marker', a DNA sequence which lies close to the gene mutation and is almost always inherited with it. Thus the presence of a linked genetic marker, identified using a marker probe, only infers the presence of the disease-linked gene mutation. Similarly, in most instances the restriction site mutations used in RFLP analysis are linked genetic markers; that is, they are rare mutations that are so close to a mutant gene

that is being screened for that there is a high probability that their presence marks the presence in the genome of the mutant gene.

The risk that the use of linked genetic markers in genetic screening can give rise to false results has caused controversy concerning their use. For example, there is controversy over the use of the G8 probe, which is complementary to a linked genetic marker called G8, which lies close to a gene mutation associated with Huntington's chorea. Huntington's chorea is a rare single-gene dominant disorder which is expressed in an individual who has inherited a mutant gene from just one parent, and will be transmitted, on average, to half of the offspring of an affected individual. Presently untreatable, it is a late-onset condition which affects the victim both physically and mentally. It is characterized by involuntary muscular movement and mental deterioration, including loss of speech, with death occurring at around the age of 55. Onset occurs at approximately 35 years, and thus after the average age of reproduction.

The use of probes in genetic screening for a genetic disorder requires a detailed knowledge of the genetic disorder because a given disorder may result from different gene mutations in different cases. The G8 linked genetic marker is inherited with the gene mutation associated with Huntington's chorea in only 95 to 99 per cent of Huntington's chorea cases, and it is not known whether all cases of Huntington's chorea are associated with the gene mutation that lies near to the G8 genetic marker. For these reasons James Gusella, the leader of the research team from the Massachussetts General Hospital in Boston that discovered the proximity of the G8 genetic marker to the gene mutation on chromosome 4 that is associated with Huntington's chorea, refused to supply copies of the G8 probe to the Medical Genetics Unit of the Oxford Regional Health Authority in Britain. The latter who wished to undertake clinical trials to test the accuracy of the probe in diagnosing the disease in patients from families with a history of Huntington's chorea (Connor 1986).

Were a wholly reliable genetic screening test to identify individuals with Huntington's chorea available, application of the screening test would benefit those individuals screened and found to be unaffected by the disease. If the test were applied prenatally with the option of selective therapeutic abortion of affected foetuses, it would enable those with the condition to take action to avoid having children with the deleterious gene mutation. However, early positive identification of individuals with the condition would replace anxiety or ignorance regarding the future, with the certainty that theirs will be a future in which they will become demented at about 35, then degenerate, and subsequently die within the next fifteen years. If this information became publicly known, the individual sufferer of the disease might experience employment discrimination and other social sanctions.

When using genetic screening techniques based on DNA sequence analysis it is important to be aware that DNA is only one component in the complex molecular structure of organisms. Changes in DNA sequence may be related to states of health, but the majority of DNA changes will be of no consequence at all to the organism, and most diseases have nothing to do with changes in a person's DNA. Single-gene disorders are the result of mutations in a gene which affect the structure or function of the protein molecule which it codes for. However, even for single-gene disorders, the individual mutant gene primarily responsible for the condition does not operate in a vacuum but against the background of the rest of the genome. The expression of a given trait may be highly variable from one person to another, a characteristic which has been especially noted for some dominant single-gene disorders.

Direct DNA analysis is prone to error. One source of error is that some common genetic disorders may result from different mutations in different families. Another source of error is over-confidence in the amount of information which can be derived from DNA. Once the underlying basis for conditions is worked out, it may be more reliable to concentrate on developing simple and cheap biochemical and monoclonal antibody screening techniques rather than to screen DNA using microgenetic analysis.

Inaccuracies in the diagnosis of genetic disorders have serious consequences. When screening for medical intervention, a false positive result may lead to inappropriate and possibly undesirable therapy being administered to a healthy individual. A false negative result would lead to failure to treat an individual in need of therapy. When screening for reproductive information, a false negative result could lead to the unexpected birth of an affected child. A false positive result could lead to decisions to terminate a pregnancy unnecessarily, adopt or remain childless, employ the use of artificial insemination or egg donation. There is also the effect on an individual of being falsely labelled, with all the consequences regarding his or her relationships and career prospects.

Clearly, the implications of mis-screening are so grave that every endeavour should be made to ensure that its incidence is minimized if not eradicated. The degree to which each test is unreliable and the nature of the relationship between the test result and the genetic constitution of the individual should be made public. It is also important that the matter of who is legally liable for a genetic mis-screening should be determined (Reilly 1977).

10.2 GENETIC DETERMINISM

Researchers are using gene probes to look for linked genetic markers which indicate a predisposition to develop common multifactorial disorders, such as Alzheimers disease, manic-depression, malignant melanoma and breast cancer, which are the result of a complex of genetic and environmental factors. However, claims that through genetic screening it is possible to identify individuals susceptible to multifactorial illnesses must be treated with caution.

It is claimed that it is possible to screen for individuals who are susceptible to particular disorders by using genetic markers of the human leucocyte antigen (HLA) system. Leucocytes are a type of white blood cell and and are part of the body's immune system. It has been found that there are antigens on the surface of human leucocytes. These are called the human leucocyte antigens (HLA), or the HLA system, and the set of genes which code for them is called the HLA complex. Just as individuals can be classified on the basis of their ABO blood group, so they can be classified using monoclonal antibodies on the basis of their HLA type. It has been found that among individuals with certain illnesses, including meningitis, acute appendicitis, influenza, psoriasis, multiple sclerosis and rheumatoid arthritis, the distribution of particular HLA types is not typical of the racial population as a whole. For example, 90 per cent of Caucasian sufferers of an arthritic condition of the spine called ankylosing spondylitis have an HLA antigen called HLA-B27 compared to 8 per cent of healthy Caucasians (Bodmer and Bodmer 1978).

Positive correlations between HLA types and illnesses should be used with caution in genetic screening because most of the associations have been found in retrospective studies, that is in studies of individuals with the illnesses. If the HLA disease association were being studied in a disease which had a significant mortality, then only surviving patients might have been included in the study (Dick 1978). Positive associations between HLA types and malignancies could indicate higher-than-average resistance resulting in increased longevity rather than susceptibility. Associations between HLA types and diseases should be investigated through prospective studies in which large numbers of healthy individuals are classified according to their HLA type and then monitored for which disorders they develop.

Using restriction enzyme analysis it will be possible to screen for what appear to be hereditary disorders even when the genetic and biochemical basis of such disorders is unknown. This apparent advantage of the use of restriction enzyme analysis in genetic screening is also its drawback. Many disorders which appear to run in families are the result of the interaction of a number of genetic and environmental factors. Whilst the genetic basis of such disorders remains a mystery, the genetic markers used in restriction enzyme analysis, are, by comparison, inherited in a simple Mendelian fashion, and it may be possible in certain population groups to identify a linked genetic marker, that of itself is not implicated in a multifactorial illness, but which none the less has a positive correlation with it. A healthy scepticism should obtain when using direct DNA analysis for the purpose of diagnosing multifactorial disorders, particularly where there is no known connection between DNA and the disease. Prospective studies, involving a large sample of randomly selected people, conducted over their entire life-spans and taking into account relevant factors of their age, race, sex and life-style, including diet, exercise, housing, occupation and environmental hazards, are highly desirable prior to the introduction of genetic screening to predict susceptibility to multifactorial disorders, because such disorders occur as a result of a large number of factors in the genetic constitution of the individual and indeterminate environmental influences (see Bodmer and Bodmer 1978).

Research has also been undertaken to establish correlations between the genetic constitution and predisposition to intelligence, antisocial behaviour, mental illness and workplace-associated disorders.

The study of the hereditability of human intelligence is a special branch of biometrical genetics. Founded by Francis Galton at the turn of the century, biometry is the application of a quantitative approach to biological problems. In the belief that an understanding of the laws of inheritance would emerge, the biometricians applied their skills to traits such as size, strength, vigour and intelligence. When Albert Binet died in 1911, the Galtonian eugenicists, believing that the intelligence test measured an innate and unchangeable quantity fixed by genetic inheritance, took control of the mental testing movement in English-speaking countries.

Studies in more recent times have attempted to use scientific techniques to show that particular human groups or populations are innately more intelligent than others. Jensen (1969), an educational psychologist, concluded that the failure of certain 'compensatory education' programmes in the USA was due to the innate inferiority of the groups for which such schemes were devised. Eysenck (1971) supported Jensen's work, and argued that groups such as blacks and Irish are relatively genetically inferior. Jensen,

Eysenck and others believe that the results of IQ testing indicate that a genetic meritocracy exists, and that variance between individuals cannot be accounted for by environmental factors.

The belief that intelligence is largely genetically determined relies on the thesis that the number of 'intelligence genes' is lower on average in certain groups of people. There is, however, no such thing as a 'low IQ gene' or a 'high IQ gene'; at best there may be particular combinations of genes which, in particular environments, produce a 'high' or 'low' IQ (S. Rose 1976; Gould 1981). In contradiction to the studies of Eysenck and Jensen, it has been found that children reared by the same mother are found to resemble her in IQ to the same degree, whether or not they share her genes (see S. Rose *et al.* 1984).

There is a direct link between the criminal anthropology of the 1890s and criminal cytogenetics. Cytogenetics is that part of genetics that pertains to the structure of chromosomes in cells. In the mid-1950s genetic screening research to correlate unusual karyotypes (chromosomal arrangements) with pathological behaviour was undertaken. A finding, to which much importance was attached, was that secure institutions had a higher proportion of males with the karyotype XYY, as distinct from the normal male XY sex chromosomal constitution, than did non-secure institutions. Some scientists began to wonder if the presence of an extra Y chromosome predisposes its carrier to unusually aggressive behaviour, and genetically determined criminality was hypothesized (Jacobs *et al.* 1967; Nielsen 1968). There was a flurry of enthusiasm for such claims in the late 1960s and early 1970s. W. Court Brown suggested that finding the apparent correlation between the XYY karyotype and disturbances of behaviour 'may be the most important discovery yet made in human cytogenetics [because] it may provide a powerful lever to open up the study of human behavioural genetics' (Court Brown 1967).

The above hypothesis was expounded despite the fact that the exact frequency of the XYY male in the general population is not known with any certainty. There is, in addition, no convincing evidence that the presence of an extra Y chromosome retards intellectual development, and indeed some XYY males are of 'normal' intelligence. Later surveys have found less difference between the frequency of XYY males in the two institutional types. XYY males in institutions tend to have depressed IQ levels and be unusually tall, and it is this latter characteristic, rather than aggressive behaviour, which has predisposed the authorities to place many of them in special security institutions instead of normal institutions (McKusick 1969).

Genetic determinism is an over-reductionist approach which assumes the characteristics of the phenotype to be innate and essentially unchangeable (see Gould 1981). The social implications of deterministic theories may be illustrated by the hypothetical case of a prenatal genetic screening test which reveals that the embryo is of a genetic constitution which, it is believed, would predispose the person to criminal behaviour. In a society which accepts and believes in the genetic determination of behaviour, then, at the very least, such a person would be born a labelled criminal; termination of the pregnancy might be considered by the family or even by society as a reasonable, if not compulsory, course of action. Whilst chemotherapy, social conditioning and physical restraint might protect society from the 'inevitable' criminal tendencies of the genetically determined deviant, the only 'cure' for his or her condition would be gene therapy. While such social policies may seem unlikely in pluralist democracies, the reader is reminded of the enthusiasm with which some well-meaning reformers in the 1960s advocated the early identification of individuals of an antisocial predisposition as suitable cases for 'treatment', the nature of which was to be decided by 'panels of experts' in order that the subject be rendered 'normal' or at least harmless (e.g. see Wootton 1963).

10.3   GENETIC SCREENING IN INDUSTRY

The development of genetic screening tests for an expanding range of traits foreshadows the steady increase of the use of genetic screening tests in industry. There are two ways in which genetic screening tests are likely to be used in the workplace, namely cytogenetic monitoring and pre-employment genetic screening.

Cytogenetic monitoring involves various ways of examining genetic changes within people's cells that may have been brought about by environmental exposures to chemicals or radiation. Modern manufacturing processes use in total over 3,500 chemical substances that do not normally occur in nature. Each year in the UK, for example, about 2,000 people are believed to die from exposure to toxic substances at work. For example, companies that produce electronic equipment and scientific instruments require employees to handle carcinogenic chemicals such as solder, asbestos and alum; companies manufacturing metal products often require employees to handle nickel, lead, solvents and chromic acid; and the petroleum industry exposes workers to benzene, naphthalene and aromatic hydrocarbons. Each of these substances creates an environment for workers in which the risk of 'unmasking' cancer is greatly increased (see Harsanyi and Hutton 1982). In addition to the deaths from known causes, there are many unexplained cancers which may be linked to exposure to hazards at work (Ehrlichman 1984).

Whilst some changes in DNA sequence may have implications for states of health and disease, most DNA changes are of little consequence to the organism. Indeed, most diseases have nothing at all to do with changes in a person's DNA (Newman 1985). Although it may be possible to trace the development of changes in DNA sequence in the genetic constitution of workers by using cytogenetic monitoring, and although workers can die as a result of exposure to toxic substances in the workplace, there is no evidence to support the case that detectable changes in DNA sequences are associated with environment-induced worker morbidity. Hence, cytogenetic monitoring is not a substitute for safe working conditions and practices.

The emphasis in industrial screening is currently on pre-employment screening for inherited genetic traits rather than testing for detrimental cytogenetic changes which may be induced by workplace conditions. In pre-employment genetic screening there is a conflict between the interests of employers, who may not want to hire or promote a person if they believe that he or she is likely to develop a debilitating disease, and the protection of the privacy of employees.

Many of the large industrial corporations in industrialized countries perform pre-employment genetic screening tests for the purpose of identifying job applicants who have inherited a trait which is believed to render them susceptible to illness triggered by specific workplace conditions. For example, it was suggested that workers with a deficiency of an enzyme called alpha-1-antitrypsin are predisposed to lung disease, and consequently would be at risk in jobs which require exposure to asbestos or cotton dust.

Some industrial chemicals induce two types of reactions in workers: acute reactions which are noticed soon after exposure; and slowly developing chronic diseases such as cancer. Some people develop acute, allergic respiratory or skin reactions to formaldehyde, for example, and cannot tolerate even the smallest amount of exposure. Other people experience mild respiratory and skin irritations. Yet there is increasing evidence that formaldehyde is a human carcinogen. Even if companies screened 'hypersusceptibles' for the allergic reactions, the remaining workers may still be at risk of developing cancer. Clearly, it is preferable to use a substitute for formaldehyde and lessen the incidence of both kinds of problems for all workers (Henifin and Hubbard 1983). Ideally, standards of health and safety should be developed for each hazard that would ensure maximum protection for the most susceptible worker irrespective of genetic status, age or sex.

Where reliable data are available on the association between the workplace environment and an occupation-induced disease, then, as John Elkington observes, employers have at least three options. They can choose to clean up the working environment. They can brief their workers on the risks and leave decisions to the individuals concerned. Or they can deny employment to those identified through genetic screening to be 'excessively' susceptible (Elkington 1983).

Excluding one set of people from a potentially hazardous occupation may be seen as welcome 'protective exclusion' but there are obvious dangers where negative job discrimination for genetic reasons is socially accepted. Because there are remarkable differences in the frequency of certain single-gene disorders between different racial groups, exclusions may run along racial and ethnic lines, reinforcing the obstacles facing those who already suffer from job discrimination. For example, carriers of sickle-cell anaemia are able to be identified in genetic screening tests through the presence of a genetic marker. On average only one in a 1,000 Anglo-Saxons carries the marker, but it appears in more than 10 per cent of Filipinos, US blacks and Mediterranean Jews.

Sickle-cell anaemia is a recessive single-gene disorder which affects the oxygen-carrying haemoglobin molecules in red blood cells. Individuals with sickle-cell anaemia have inherited a mutant gene from both parents, and all of their haemoglobin molecules are of an abnormal type called sickle haemoglobin, or HbS. When red blood cells containing sickle haemoglobin molecules lose oxygen, they can become crescent (sickle) shaped, rather than disc-shaped, and clump together, thus failing to circulate properly through small blood vessels. This can lead to severe tissue damage and infection, joint pains, incapacitation and even death. Individuals with sickle-cell anaemia are almost always too ill to work in an industrial environment.

Carriers of sickle-cell anaemia have inherited one normal haemoglobin gene and one sickle haemoglobin gene and have a condition known as sickle-cell trait. Between 20 and 40 per cent of the red blood cells of individuals with sickle-cell trait are affected. Such carriers have minimal clinical problems; their average life expectancy is not reduced, nor are they more often in hospital than individuals with normal haemoglobin. Only a major decrease in the oxygen concentration causes their red blood cells to sickle (Harsanyi and Hutton 1983).

Some doctors have suggested that people with the sickle-cell marker, that is individuals with sickle-cell trait, should be warned against exposing themselves to sudden drops in oxygen concentration as can occur in mine rescue work or high-altitude flying. As a consequence of exaggerated fears arising from this information, coupled with the ability to identify carrier individuals, job opportunities have been lost; black airline employees in the USA were at one time grounded because of fears that they might go into a 'sickling crisis' if the aeroplane depressurized: insurance companies, without adequate supporting data, have charged higher insurance premiums to carriers; and carriers in the USA have even been deferred by the armed forces (see Harsanyi and Hutton 1982).

Industrial health science has concentrated on the individual worker and not on the technological environment in which he or she is placed. The tendency has been to attribute blame for the incidence of accidents and ill health at work to the individual employee: the individual is blamed for apathy towards safety, accident proneness, genetic susceptibility and carelessness concerning protective clothing. Genetic screening can be used to shift responsibility for work-related diseases onto the workers. This is especially risky for minorities who are already at a disadvantage in the job market. Occupational disease can be prevented by reducing workplace exposure for all workers rather than by pre-employment genetic screening for so-called susceptible individuals. In short, genetic screening in the workplace has both the potential to protect the health of workers and to be used to cover up for a combination of inadequate workplace conditions and worker discrimination.

10.4 EUGENICS AND THE NEW REPRODUCTIVE TECHNOLOGIES

The popular interpretation of Darwin's thesis, that natural selection is the process that gives direction to evolution, propounded in his famous book, *On the Origin of Species*, finally published in 1859, provided the impetus for modern eugenics both as a science and as a social movement. Eugenics is the science of the influences that improve the inborn qualities of the human race and develop them to the utmost advantage. As a social movement, eugenics encompasses all efforts whose goal is to bring about genetic change in a particular direction within given populations or to the human species as a whole. Ten years after the publication of Darwin's theory of evolution, Sir Francis Galton, Charles Darwin's cousin, in his book *Hereditary Genius* argued that our abilities are derived by inheritance. 'I conclude', he wrote, 'that each generation has enormous power over the natural gifts of those that follow, and maintain that it is a duty that we owe to humanity to investigate the range of that power, and to exercise it in such a way that, without being unwise towards ourselves, shall be most advantageous to future inhabitants of the earth' (Galton 1869). In 1883 Galton coined the term 'eugenics', which he defined as 'the study of agencies under social control that may improve or impair . . . future generations physically or mentally' (Galton 1883).

Microgenetic engineering and the new reproductive technologies enormously enhance the potential prospects for eugenics.

Microgenetic engineering of cells of the reproductive system, cloning, birth control, *in vitro* fertilization (IVF), genetic screening, artificial insemination (AI), egg donation and cryobiology (the techniques to freeze and thaw biological entities without impairing their function) theoretically extend the potential for the eugenic manipulation of the genetic constitution of the human race.

The new reproductive technologies are artificial insemination, egg donation, *in vitro* fertilization and embryo transfer. Artificial insemination is the application of sperm to an unfertilized egg by means other than ejaculation during sexual intercourse. Artificial insemination by a donor (AID) is the fertilization of an egg using sperm donated by a third party to a woman wishing to conceive a child.

Egg donation is the donation of unfertilized eggs by one female to another. The first stage in the procedure is to administer fertility-drug therapy to the donor. This stimulates her ovaries to release several eggs. The eggs are then removed through a fine tube which is inserted through her abdominal wall.

In vitro, meaning in glass, is, as we have mentioned, a scientific term used in contrast with the term *in vivo*, meaning in the living organism. *In vitro* fertilization (IVF) is the fertilization of extracted eggs by mixing them with sperm in a glass dish (a petri dish) rather than in the reproductive system of a living female. So-called 'test-tube babies' are the products of IVF. After fertilization *in vitro*, the eggs are carefully examined to ensure their cells are dividing normally, and two or three of those which are found to be normal are transferred, or implanted, into the uterus of the female who is to be the gestator.

The genetic constitution of the sperm, eggs and IVF eggs are each potential targets for genetic screening. However, the very process of genetic screening may cause damage to the sperm and eggs, resulting in increased levels of abnormality and miscarriage. Also, if the screening of eggs fertilized *in vitro* is to be practicable, a high proportion of screened eggs must be capable of successful development. At present only about 10 per cent of IVF eggs which are implanted result in viable pregnancies.

Figure 10.4.1 illustrates how AID and egg donation extend the pool of genetic material available to a couple wishing to have a child. In addition to the more common combination of the fertilization of the egg of the female by the sperm of her mate, represented by A on Figure 10.4.1, the couple may use the mate's sperm with a donor's egg (B), the female's egg may be fertilized by a donor's sperm (C), or a donor's egg may be fertilized by a donor's sperm (D). The fertilized egg may be brought to term in the uterus of the female (E), the egg donor (F) or a surrogate gestator (G). The genetic constitution of the gestator does not, of course, affect the genetic constitution of the fertilized egg.

Approximately one in ten couples have difficulty in conceiving a child, and from this market alone, despite legal constraints in many countries, the annual market value for surrogacy exceeds one million dollars. There is no worldwide consensus on the regulation of surrogacy arrangements and the rights to babies born of surrogate mothers are determined by several different principles in different

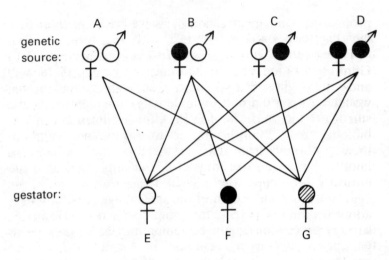

Key:

couple (♀ female, ♂ male) who wish to have a child

♀ egg donor (who may also be the gestator)

♂ sperm donor

♀ surrogate gestator

Figure 10.4.1 Artificial insemination, egg donation and surrogacy schema

countries. In the USA, the rights to the baby are determined by contractual arrangements made between the surrogate mother and the would-be adoptive parents. In West Germany, legal parentage is conferred on the woman who gives birth to the child, regardless of genetic relatedness. Other criteria could be to award the child to those adults who donated the egg and sperm, or those deemed best in the child's interest, which could be a welfare agency.

In the USA, where there is a *laissez faire* policy towards surrogacy, there are a number of commercial surrogacy agencies, for example, the Infertility Centre of Michigan (ICM). The New York office of ICM was involved in the adoption of 'Baby M' in New Jersey in 1986. The New Jersey courts set a US precedent when they awarded custody of Baby M to the biological father and his wife despite the protests of the surrogate mother who bore the child for a fee (Dickman 1987)[1].

This contrasts with the legal position of surrogacy in West Germany, where surrogacy contracts are considered null and void. Under the adoption law, West German courts would award a baby born to a surrogate mother to the surrogate, even if she had signed a contract with the biological father and his partner to surrender the baby to them. Proposals have been made to change the adoption laws to strengthen the position of the surrogate mother by ensuring she keeps the baby even when both sperm and egg, fertilized *in vitro*, have come from the would-be adoptive parents (Dickman 1987). In Britain, commercial surrogacy arrangements are banned under the 1985 Surrogacy Arrangements Act. Under this Act only third parties are liable; this means that even when the arrangement is made through a commercial agency the surrogate mother and commissioning parents are free from prosecution (Yoxen 1986). Were the recommendations of a White Paper published in Britain in November 1987 to be enacted, the parental rights over a surrogate child in Britain would be similar to those in West Germany. The White Paper proposes that any surrogacy contracts will be unenforceable in the British courts and that surrogate mothers will have all legal rights over the child (*Human Fertilisation and Embryology: A Framework for Legislation* 1987). In France surrogacy agencies were outlawed in 1987 under clause 351 of the French civil law which states that it is an offence for an individual to incite a woman to enter into an agreement to abandon her child.

As Erwin Chargaff has observed, human husbandry is so new a profession that society has not yet learnt how to protect itself (Chargaff 1987). Difficulties which can arise where human sex cells and embryos are stored for potential use include the difficulty of tracing donors and the question of rights of succession, for example, in a case where a woman wishes to be inseminated with her late husband's sperm, or if she wished to have a stored embryo transferred to her womb. The British White Paper published in November 1987 proposed that there should be a limit of five years on storage of frozen embryos and that embryos should not be implanted into another woman, nor used for research, nor destroyed without the consent of both donors.

Positive eugenics is the assigning of members of the population to parentage. In Plato's utopian *Republic*, for example, union between the 'better' specimens of both sexes was encouraged. More recently, in 1985 Singapore's Prime Minister, Lee Kuan Yew, initiated a programme of social events to encourage more graduate women to marry and reproduce with the aim of increasing the intelligence stock of Singapore (Cater 1985). Hermann Muller, who received a Nobel Prize in 1946 for his work on the mutagenic property of X-rays, believed that artificial insemination should be used for positive eugenic purposes (Muller 1935, 1963). This practice, called 'eutelegenesis', a term coined by Herbert Brewer, would employ sperm of a donor chosen on the basis of eugenic criteria (Brewer 1935). Modern reproductive technology permits the incorporation of the eggs of what are considered eugenically desirable females into such a scheme.

A clone is a number of identical molecules, cells or organisms. A clone of cells or organisms is derived from a single cell or organism by asexual multiplication in such a way that each member of the clone is genetically identical with every other member and with the single cell or organism from which the clone is derived. Cutting, grafting and splitting tubers produce clones of plants and are methods of propagation exploited in plant breeding to obtain identical copies of what is perceived of as genetically desirable specimens.

Positive eugenic scenarios sometimes envisage the replication of the entire genetic constitution of what is considered to be a supremely eugenically desirable individual through cloning. Cloning of higher animals requires 'nuclear transplantation'. To make a clone—a genetically identical copy—of a higher animal, the nucleus of one of its cells must be transplanted into a fertilized egg from which the nucleus has been removed (i.e. enucleated); this is depicted in Figure 10.4.2. The fertilized egg with the transplanted nucleus is cultured *in vitro* before being implanted into the uterus (womb) of a female—a surrogate gestator—who will nurture the cloned embryo until parturition (birth).

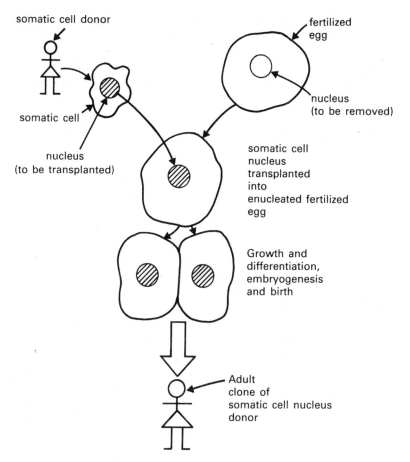

Figure 10.4.2 Cloning

Nuclear transplantation in animals has mostly been attempted on amphibians, for example frogs, because their fertilized eggs (spawn) are large and easy to obtain (Gurdon 1977). The cloning of adult mammals (including humans) through nuclear transplantation is technically more complicated than that of amphibians, and although claims to have successfully cloned mice have been made, such experiments have not been repeated and independently confirmed (see Illmensee and Hoppe 1981; Marx 1981).

The motivation for advocating positive eugenic policies is the desire to improve the performance of a given population with respect to what are perceived to be desirable faculties and the belief that this can be attained through the manipulation of the genetic constitution of the population. However, it is not usually a shortage of human potential that is the limiting factor in facilitating achievement by a given population, but the paucity of social environments conducive to the full development of the human potential that resides therein. For example, if it is considered desirable that a society should have more creative people, then it is more economical to direct scarce resources to provide further facilities in appropriate areas, such as the arts, humanities and sciences, rather than to gamble such resources on the chance that one such desirable individual might be 'genetically fixed' and viable. Furthermore, as Sir Peter Medawar (1974) argued, whether desirable or not, a programme of positive eugenics is not feasible owing to the present ignorance of how to achieve a stated eugenic goal in precise terms of human genetics.

In the nineteenth-century many social reformers held the view that a eugenic distribution of births should be achieved primarily through birth control. Neo-Malthusians, so-called because their views were derived from the theory of population propounded by Thomas Malthus, believed that the population growth of the lower orders of nineteenth-century industrial society was leading to a deterioration of the genetic quality of the population. They proposed negative eugenics—social programmes designed to reduce the birth rate of what they considered to be the 'less fit' members of the population.

Figure 10.4.3 outlines a hypothetical schema for modern negative eugenics. According to Reilly (1977), a socially conscientious system would include a national registry of genetic profiles derived from genetic screening tests performed routinely at birth and fed into a computer data bank. A preliminary scan of the genetic profile would identify individuals requiring immediate therapy. When a couple applied for a marriage licence their respective genetic profiles would be analysed to reveal which genetic traits they were both carriers of, in order to identify which genetic disorders, if any, their children would be at risk of being afflicted with. If it were discovered that there was a strong likelihood that their offspring would inherit what was considered to be an undesirable genetic constitution, they could utilize the eggs or sperm of a genetically screened donor as alternative sources of genetic material as illustrated in Figure 10.4.3a.

Figure 10.4.3b illustrates the role of prenatal genetic screening in modern negative eugenics. Major congenital malformations can be detected prenatally using X-ray or ultrasound analysis. Foetal blood sampling enables the prenatal diagnosis of genetic blood disorders, such as sickle-cell anaemia, the thalassaemias and haemophilia, and of disorders in which there are altered levels of metabolites in the blood. Amniocentesis is the removal of amniotic fluid and floating cells which surround the foetus within the amniotic sac (water sac) inside the uterus (womb). Both

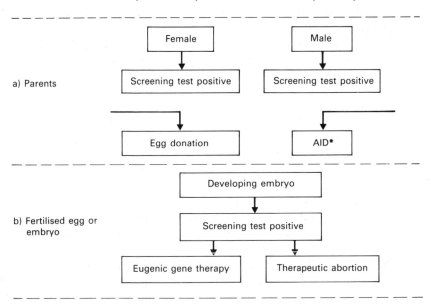

Note: *AID: Artificial Insemination by Donor*

Figure 10.4.3 Potential role of reproductive engineering, gene therapy and genetic screening in negative eugenics

the fluid and the cells can be analysed. Examination of the amniotic fluid is used mainly as an indicator of neural tube defects, such as spina bifida and anencephaly, which have an incidence of between one and two per 1,000 live births. Studies of the chromosomes in the foetal cells can identify the sex of the foetus and determine the presence of chromosomal aberrations. Biochemical tests on the cultured cells to measure enzyme levels are able to diagnose many single-gene disorders which alter the metabolism, including Lesch-Nyhan syndrome and adenosine deaminase (ADA) deficiency.

Amniocentesis and foetal blood sampling cannot be conducted until the second trimester of the pregnancy and the analysis may not be completed until late in the second trimester. Chorionic villus sampling (CVS) is a recently developed prenatal diagnostic technique which can be performed as early as the eighth or ninth week of pregnancy enabling therapeutic termination of the pregnancy at an earlier stage with less risk to the mother. CVS is the direct sampling of cells of the chorion, the membrane which surrounds the human embryo. The cells of the chorion are derived from the fertilized egg which develops into the foetus, and contain the same genetic information as those of the foetus. Therefore the chorion is a source of cells containing foetal DNA which can be screened using microgenetic analysis as a proxy for the genetic screening of the foetus. Chorionic tissue can also be used for chromosomal and biochemical analysis and provides an opportunity to diagnose genetic disorders within the first trimester of pregnancy.

Were a prenatal genetic screening test to identify an embryo afflicted with what was considered to be an undesirable genetic trait, the birth of an affected individual could be prevented by the therapeutic abortion of the embryo (Singer and Wells 1984) or, more futuristically, averted by attempting eugenic gene therapy, as shown in Figure 10.4.3b. Microgenetic manipulation of the germ cells and the newly fertilized egg, if successful, would be eugenic because the genetic modification would be present in the germ-line of the recipient and may be transmitted to future descendents, thus altering the genetic constitution of the population (see Fig. 9.1.1).

Negative eugenics as a socio-political movement flourished from the late nineteenth-century to the 1930s. During this period ignorance and misplaced utopian ideology have conspired to produce many inhumane proposals. In the early 1890s, for example, there were sporadic reports of mass castrations of retarded children conducted by institutions. In 1917 some sixteen states of the USA had compulsory sterilization laws for the 'feebleminded', the 'insane', the 'habitual criminal' and other categories of the 'hereditary unfit'. By 1935 approximately 30,000 eugenic sterilizations had been carried out in the USA (Bajema 1976).

Negative eugenics policies such as those described above were partly inspired by the tendency to apply inappropriately the principles of particulate inheritance to disorders which are the result of a complex of genetic and environmental factors. For example, it has been argued that schizophrenia can be attributed to a recessive disorder in a single gene (Kallmann 1938). Kallmann advocated the sterilization of not only schizophrenics but of those relatives who were supposed *carriers* of the 'schizophrenia gene', in a *reductio ad absurdum* programme to purge the human race of the 'gene for schizophrenia' (see Rose *et al.* 1984).

In the unlikely event of Kallmann's hypothesis regarding the inheritance of schizophrenia being substantiated, his purgative approach is not a practicable negative eugenics policy. As Table 9.3.2 shows (see p. 228), the frequency of carriers of the recessive genes responsible for single-gene disorders is surprisingly high compared to the frequency of the disorder. If a single-gene recessive disorder were to occur with a frequency of one in 10,000 in the population, then the frequency of the deleterious gene assumed to be responsible for the disorder would be one in 100 and the frequency of its genetic carriers as high as one in 50. Complete eradication of the defective gene responsible for the condition would require the sterilization of one-fiftieth of the total population (see Medawar and Medawar 1977). Each of us carries between three and eight genes for lethal recessive disorders (Howell 1973). Assuming perfect genetic screening measures and unlimited economic resources, the implementation of a negative eugenics policy to eradicate all lethal recessive genes would threaten the very survival of the human race!

The new technologies for human reproduction have prompted heated public controversy because they raise emotive moral issues and complex legal questions.

Moral questions arise, such as, should prenatal genetic screening tests be administered for conditions for which there is no treatment and in states where legislation forbids therapeutic abortion? Choices available are constrained within a legal framework and this framework reflects the social relationship between law and medicine. For example, the medical choices in the management of genetic disease are contingent upon legislation governing abortion, euthanasia, IVF, embryo experimentation, artificial insemination and egg donation.

The discarding of those eggs which are fertilized *in vitro* but not implanted because they are imperfect, and the selective abortion of foetuses with incurable conditions has created controversy. The so-called 'IVF controversy' largely concerns the fate of those eggs and embryos not needed for reproductive purposes, some of which are used for scientific research (see Grobstein *et al.* 1983; Council for Science and Society 1984; Singer and Wells 1984; Yoxen 1986). Public anxiety about the uses that might be made in laboratories of artificially fertilized human embryos requires that research projects in this field should be regulated and that breaches of the regulations should be criminal offences. In Britain the Committee of Inquiry into Human Fertilisation and Embryology, chaired by Lady Mary Warnock, concluded in its 1984 report that IVF techniques are a welcome means of treating infertility which nevertheless prompt legal questions about parentage (*Committee of Inquiry into Human Fertilisation and Embryology* 1984). The Warnock Committee concluded that the human embryo should be protected, and because the committee considered that human embryos might have the potential for sensory perception after fourteen days after fertilization, they recommended that research on them after this stage of development should be made illegal. The Warnock Committee recommended that legislation be enacted which would establish a statutory licensing authority. In the UK the Voluntary Licensing Authority (for human IVF and embryology) believed that both its own work and research in human embryology were being hampered by the continuing lack of legislation. The Voluntary Licensing Authority was the creation of the MRC and the Royal College of Obstetricians and Gynaecologists, but the licenses it issued to research centres had no statutory force. Clearly, this is an area where there is a need for regulatory legislation and for the establishment of a statutory body with a qualified inspectorate to monitor and control all treatment and research. Interestingly, it has been proposed in a White Paper published in Britain in November 1987 that the new statutory body be chaired by someone who is not a doctor or a scientist involved in the field and that at least half the members of the body should be lay people (*Human Fertilisation and Embryology: A Framework for Legislation 1987*). This new statutory body would be responsible for licensing research on human IVF and embryology. Under the recommendations of this White Paper, it would be a criminal offence to create, clone, use or store a human embryo outside the body or to fuse animal and human cells without a permit from a new statutory licensing authority. These recommendations suggest that research on 'spare embryos' be restricted to that aimed at advances in diagnostic or therapeutic techniques or infertility control.

With the new reproductive technologies medical researchers are shifting their focus from the end of pregnancy to its beginning. The ability to diagnose and treat fertilized eggs would be a logical extension of this new research emphasis.

The genetic therapy of embryos is seen by some groups as a potential substitute for current screening tests and therapeutic abortion. If legal rights for embryos were established and abortions banned (as pressure groups such as LIFE and the SPUC wish), then the only legal means of preventing genetic diseases would be the diagnosis and treatment of embryos and foetuses. Gene therapy would be a new source of medical intervention offering possibilities for control over how and what sort of babies women give birth to. Protection of the rights of the embryo, combined with the availability of gene therapy, could even mean that pregnant women could be held to have a legal obligation to undergo embryo therapy in circumstances where the embryo was believed to be afflicted with a disorder for which therapy existed. Even in countries in which abortion is legal, women have on occasion already been brought to court by physicians for refusing to have caesarean sections (Minden 1985).

According to the philosophy of Natural Rights, the embryo, as a potential human being, has all the normal rights of an individual, including the right to inviolability. An embryo cannot consent to be experimented on, or to its own destruction. To perform research upon it, to kill it or to deny it the uterine environment necessary for its life, are therefore illicit; such an absolutist philosophy of the 'sanctity of life' knows of no exceptions. However, just as an embryo cannot consent to be experimented upon, neither can it consent to be operated upon surgically. Moreover, there is no consensus over what exactly constitutes the beginning of the life of an individual. Biologically, life is a continuum with no discrete beginning or end between generations (see Mackenzie 1984). The genome, the unique genetic constitution, of an individual is first formed at fertilization. However, in the view of some Christians the embryo gains the status of a human being at between forty and eighty days after fertilization, when it is unmistakeably recognizable as a member of the human species (Dunstan 1984). The ethic of the 'sanctity of life' is being challenged by the desire for perfectability. John Stuart Mill believed that we should 'amend the course of nature'. In his famous *Essay on Liberty* Mill wrote, 'To bestow a life which may be either a curse or a blessing, unless the being on whom it is to be bestowed will have at least the ordinary chances of a desirable existence, is a crime against that being' (Mill 1859). According to this philosophy of humanitarian utilitarianism, parents would have the right to choose to prevent the birth of their progeny known to be afflicted with serious and effectively untreatable genetic disorders.

10.5 THE GRAIL OF HUMAN GENETICS

The grail of human genetics—the quest to sequence all the genetic material characteristic of the human species—will undoubtedly result in the development of technology which will increase the speed and ease with which the entire genetic material of individuals can be decoded. Computer technology will enable the scanning of each genetic profile, comprising a long sequence of bases, in search of permutations and combinations of bases believed to be associated with or responsible for phenotypic traits.

DNA fingerprinting is a technique which uses gene probes to generate personal genetic profiles which are as specific to individuals as conventional fingerprints. Throughout the DNA of the human genome there are regions, called hypervariable regions, which exhibit a high degree of variability in length from individual to individual. These hypervariable regions are comprised of the so-called minisatellites which consist of a short DNA sequence repeated a number of times in tandem. The number of repeats in a minisatellite at a given location in the genome varies between people and is inherited, and it is this minisatellite variation which is exploited in DNA fingerprinting. DNA fingerprinting uses genetic probes based on minisatellite DNA sequences to detect individual variation in hypervariable regions of the genome (Jeffreys *et al.* 1985a).

It is envisaged that DNA fingerprinting will revolutionize forensic biology. Biological samples for forensic analysis are typically bloodstains or semen stains on cloth or other surfaces, often several days or even weeks old, vaginal swabs taken after an alleged rape and sometimes hair roots. DNA fingerprints obtained from bloodstains that were four years old, hair roots and even semen stains have all been shown to be specific to individuals when compared with whole blood and semen samples. Also, the minisatellite probes used in DNA fingerprinting provide a method for paternity and maternity testing which is used both to prove and to disprove parenthood. They have been patented and DNA fingerprinting tests marketed by ICI Cellmark Diagnostics are being used by the police force and the Home Office in Britain (Jeffreys *et al.* 1985a). They have already been used to screen suspects in rape cases and to resolve immigration disputes arising from lack of proof of family relationships (Jeffreys *et al.* 1985b). Alec Jeffreys of Leicester University, England, who invented DNA fingerprinting, has urged that it be used with caution in pedigree analysis. DNA fingerprinting is only accurate in determining relationships in approximately 95 per cent of cases. A DNA fingerprint difference between a child and its two claimed parents may be the result of a new mutation in the germ-line rather than a false claim of parentage.

Employers, insurance companies, educational establishments, the police, the law courts and immigration authorities are interested in collecting the respective genetic profiles of potential employees, policy holders, students, criminal suspects, paternity defendants and immigrants. The world's first computerized data bank of DNA fingerprint information on convicted criminals is being planned for use in the USA by 1990. The ability to record the genetic profile of an individual raises the issue of confidentiality. In the USA the majority of states do not have statutes that recognize the confidentiality of public health information. Life insurance companies have traditionally excluded people from insurance protection because their health is poor or because they are at high risk of becoming seriously ill. With respect to employment, this information may be used in a number of ways which may benefit, and conversely, may be detrimental to, different individuals or groups (Lappe 1985).

The AIDS screening test, whilst not a genetic screening test, illustrates the dilemma of under what circumstances a screening test should be compulsory; which agents in society should be authorized to conduct such a screening test; and which persons or agencies should have access to the results of such a test. The psychological, political and moral dangers of legislating for human genetic screening, whether for predictive, diagnostic or identification purposes, may sometimes outweigh the potential medical and economic benefits. Third-party access to genetic screening data has potentially detrimental consequences for the career and personal relationships of the individual screened. Thus prerequisites for genetic screening programmes should include well-defined procedures for obtaining informed consent and safeguards for protecting the rights of individual and family privacy.

The relationship between genotype and behaviour, personality, intelligence and susceptibility to illness are not deterministic. Even for single-gene disorders, there is a need for careful questioning of the validity of claimed correlations between genetic markers and phenotypes. Whilst genetic screening has the potential to be of great benefit to particular individuals and to our social welfare, constant vigilance is needed lest genetic screening be used for smuggling into society illiberal forms of social engineering and as a means whereby individual freedoms can be abused. It is our responsibility to ensure that genetic screening does not become the covert vehicle for instigating policies of racial discrimination or a justification for reducing compensatory educational programmes, or an excuse for not making the work environment safe for all workers.

NOTE

[1] In February 1988 the commercial future of surrogacy agencies in the USA became uncertain when the New Jersey Supreme Court outlawed commercial surrogate motherhood contracts.

Epilogue

Genetic engineering is in grave danger of becoming an imprudent technology. Society should pause and take stock of the dangers inherent in the genetic engineering enterprise, lest we all suffer a similar catastrophe to that of Icarus whose imprudent use of technology proved fatal.

The gap between the Daedalean power of this revolutionary new science and technology and our inability to foresee it consequences creates a moral duty which can only be executed when the utopian ideal of perfectability, which is embedded in the scientific endeavour, is superseded by one of greater responsibility and accountability.

'Ignorance is the parent of fear', Herman Melville wrote in his novel *Moby Dick*. In the field of genetic engineering, ill-informed discussion may lead, on the one hand, to the prohibition of desirable and efficacious biological innovations, or, on the other, to the burgeoning of irresponsible or nefarious science with catastrophic implications. Hence there is an urgent need to increase worldwide awareness of the social, economic and ethical implications raised by the applications of modern genetic engineering techniques and to create an atmosphere in which responsible decisions can be taken regarding their procurement or prohibition.

The human condition is essentially one of interpretation and debate which implies political decision-making. The growth of knowledge and the pervasiveness of advanced technology with all its ramifications make it imperative that informed responsible debate of the implications of science and technology should take place. Judgements must be supported by reasons, and these reasons are subject to changes in the course of time. As Bernstein (1983) argues, it is important, however, that we overcome the 'Cartesian Anxiety' created by the belief in final and absolute knowledge. We should be humbled, but not intimidated, by the realization that we are, and are certain to remain, as Alexander Pope (1688–1744) wrote:

> Sole judge of truth,
> In endless error hurled.

UTOPIA = an imaginary place or state of things where everything is perfect.

References

Abraham, R.H. and Shaw, C.D. (1981), *Dynamics: The Geometry of Behavior* (Santa Cruz: Aerial Press).

ACGM/HSE/Note 2, *Disabled Host/Vector Systems* (London: ACGM).

ACGM/HSE/Note 3, *The Planned Release of Genetically Manipulated Organisms* (London: ACGM).

ACGM/HSE/Note 4, *Guidelines for the Health Surveillance of Those Involved in Genetic Manipulations at Laboratory and Large-Scale* (London: ACGM).

Albrecht-Buehler, G. (1978), 'The track of moving cells', *Scientific American* 238: 68–76.

Anderson, I. (1987), 'New genes cure a shivering mouse', *New Scientist*, 5 March, p. 24.

Anderson, W.F. (1984), 'Prospects for human gene therapy', *Science* 226: 401–9.

Anderson, W.F. (1985), 'Human gene therapy: scientific and ethical considerations', *Recombinant DNA Technical Bulletin* 8(2): 55–63.

Arrow, K.J. (1970), *Essays in the Theory of Risk-Bearing* (Amsterdam: North Holland).

Ashby Report (1975), *Report of the Working Party on the Experimental Manipulation of the Genetic Composition of Micro-organisms* (London: HMSO, Cmnd 5880).

Asinof, R. (1984), 'Averting genetic warfare', *Environmental Action*, June, pp. 16–23.

Bajema, C.J. (1976), *Eugenics Then and Now* (Stroudsburg, Penn.: Dowden, Hutchinson & Ross).

Barinaga, M. (1987a), 'Field test of ice-minus bacteria goes ahead despite vandals', *Nature* 326: 819.

Barinaga, M. (1987b), 'Critics denounce first genome map as premature', *Nature* 329: 571.

Bartels, D. (1984), 'Occupational hazards in oncogene research', *GeneWATCH* 1 (5,6): 6–8.

Beardsley, T. (1986), 'USDA goes too public too quickly', *Nature* 320: 473.

Bereano, P.L. (1984), 'Institutional Biosafety Committees and the inadequacies of risk regulation', *Science, Technology and Human Values* 9(4): 16–34.

Berg, P. *et al.* (1974), 'Potential biohazards of recombinant DNA molecules', *Science* 185: 303.

Berg. P. *et al.* (1975), 'Asilomar conference on recombinant DNA molecules', *Science* 188: 931–5.

Bernstein, R.J. (1983), *Beyond Objectivism and Relativism* (Oxford: Basil Blackwell).

Betz, F. *et al.* (1983), 'Safety aspects of genetically engineered microbial pesticides', *Recombinant DNA Technical Bulletin* 6(4): 135–41.

Biobusiness World Data Base (1983) (Washington DC: McGraw-Hill).

Blakeslee, S. (1985), 'Another joint venture', *Nature* 313: 261.

Bodmer, W.F. and Bodmer, J.G. (1978), 'Evolution and function of the HLA system', *British Medical Bulletin* 34(3): 309–16.

Boffey, P.M. (1985), 'U.S. approves field test of gene-altered bacteria', *The New York Times*, 15 November, p. A17.

Brewer, H. (1935) 'Eutelegenesis', *Eugenics Review* 28: 11–31.

Brill, W.J. (1985), 'Safety concerns and genetic engineering in agriculture', *Science* 227: 381–4.

Brinster, R.L. *et al.* (1983), 'Expression of a microinjected immunoglobulin gene in the spleen of transgenic mice', *Nature* 306: 332–6.

Budapest Treaty on the International Recognition of the Deposit of Microorganisms for the Purposes of Patent Procedure (With Regulations) (1981) (London: HMSO).

Bull, A.T. *et al.* (1982), *Biotechnology: International Trends and Perspectives* (Paris: OECD).

Buting, W.E. (1984), 'Issues in biotechnology litigation', *Biotech '84 USA*, pp. 39–48 (Online Publications, Pinner, UK).

Carlson, E. (1966), *The Gene: A Critical History* (Philadelphia: W.B. Saunders).

Cater, N. (1985), 'Eugenics "loveboat" heads for port', *New Scientist*, 16 May, p. 8.

CERB (1977), 'CERB Report', *The Bulletin of Atomic Scientists* 33, May: 23–7.

Chang, A.C.Y. and Cohen, S.N. (1974), 'Genome construction between bacterial species *in vitro*: replication and expression of *Staphylococcus* plasmid genes in *Escherichia coli'*, *Proceedings of the National Academy of Sciences* (USA) 71: 1033–4.

Chang, S. and Cohen, S.N. (1977), '*In vivo* site-specific genetic recombination promoted by EcoRI restriction endonuclease', *Proceedings of the National Academy of Sciences* (USA) 74: 4811–15.

Chargaff, E. (1987), 'Engineering a molecular nightmare', *Nature* 327: 199–200.

Cheah, M.S.C. *et al.* (1986), '*frg* proto-oncogene mRNA induced in B lymphocytes by Epstein-Barr virus infection', *Nature*, 319: 238–40.

Cherfas, J. (1982), *Man-Made Life* (Oxford: Blackwell).

Clark, J. *et al.* (1984), 'Long waves, inventions, and innovations', in C. Freeman (ed.), *Long Waves in the World Economy* (London: Frances Pinter).

Clarke, C.A. (1974), *Human Genetics and Medicine* (London: Edward Arnold).

Cohen, P.S. and Laux, D.C. (1985), '*E. coli* colonization of the mammalian colon: understanding the process', *Recombinant DNA Technical Bulletin* 8: 51–4.

Cohen, S. *et al.* (1973), 'Construction of biologically functional bacterial plasmids *in vitro*', *Proceedings of the National Academy of Sciences* (USA) 70: 3240–4.

Colwell, R.K. *et al.* (1985), 'Genetic engineering in agriculture, Letters, *Science* 229: 111–12.

Committee of Inquiry into Human Fertilization and Embryology (1984), (London: HMSO).

Committee on Health and Safety at Work (1972), *Safety and Health at Work: Report of the Committee* (Robens Report), Cmnd 5034 (London: HMSO).

Connor, S. (1986), 'Researcher witholds gene probe, *New Scientist*, 13 March, p. 17.

Connor, S. (1987a), 'AIDS: mystery of the missing data', *New Scientist*, 12 February, p. 19.

Connor, S. (1987b), 'AIDS: science stands on trial', *New Scientist*, 12 February, pp. 49–58.

Connor, S. (1987c), 'AIDS truce brings history to a halt', *New Scientist*, 9 April, p. 21.

Coser, L.A. (1984), *Refugee Scholars in America: Their Impact and Their Experience* (New Haven, Conn.: Yale University Press).

Council for Science and Society (1984), *Human Procreation: Ethical Aspects of the New Technologies* (Oxford: Oxford University Press).

Court Brown, W.M. (1967), *Human Population Cytogenetics* (Amsterdam: North Holland).

Crafts-Lighty, A. (1987), *Information Sources in Biotechnology* (Basingstoke: Macmillan).

Crick, F.C. (1958), 'On protein synthesis', *Symposium of the Society for Experimental Biology* 12: 138.

Cross, F. (1986), 'Lasers take a shine to medicine', *New Scientist*, 20 February, pp. 38–42.

Dancis, J. (1973), 'The prenatal detection of hereditary defects', in V.A. McKusick and R. Claiborne (eds.), *Medical Genetics* (New York: HP Publishing), pp. 246–53.

Darwin, C. and Wallace, A. (1858), 'On the tendency of species to form varieties; and on the perpetuation of varieties and species by natural means of selection', *Journal of the Proceedings of the Linnaean Society* 2 (20 Aug.): 45–62.

Darwin, C. (1859), *On the Origin of Species by Means of Natural Selection or the Preservation of Favoured Races in the Struggle for Life* (London: John Murray; reprinted, Cambridge, Mass.: Harvard University Press, 1975).

Davis, B.D. (1987), 'Bacterial domestication: underlying assumptions', *Science* 235: 1329–35.

Davis, J.C. (1984), 'The history of Utopia: the chronology of nowhere', in P. Alexander and R. Gill (eds.), *Utopias* (London: Duckworth).

Dawkins, R. (1982), *The Selfish Gene* (London: Paladin).

Day, M.J. (1982), *Plasmids* (London: Arnold).

Denselow, J. (1982), 'GMAG and the teenage jackass', *New Scientist*, 26 August, pp. 558–61.

Diamond v. Chakrabarty (1980), 100 S.Ct. 2204, 206 USPQ 193.

Dick, H. (1978), 'HLA and disease: introductory review', *British Medical Bulletin* 34 (3): 271–4.

Dickman, S. (1987), 'West German ructions over US surrogacy company', *Nature* 329: 577.

Dixon, B. (1984), 'Psychobiological warfare', *New Scientist*, 25 October, p. 40.

Doolittle, W.F. (1982), 'Selfish genes, the phenotype paradigm and genome evolution', in J. Maynard Smith (ed.), *Evolution Now: A Century After Darwin* (London: Nature, Macmillan).

Drummond, M. (1983), 'Plant genetic engineering: launching genes across phylogenetic barriers', *Nature* 303: 198–9.

Dunstan, G. (1984), 'Oh baby! What a moral dilemma', *The Times Higher Education Supplement*, 29 June, p. 15.

European Biotechnology Information Project (EBIP) (1985), *Seminar on Biotechnology Information* (London: EBIP).

Ehrlichman, J. (1984), 'Unions praise proposed reforms on work hazards', *Guardian*, 23 August, p. 3.

Elkington, J. (1983), 'Are you fit for your job?', *Guardian*, 30 June, p. 13.

Elkington, J. (1985), *The Gene Factory* (London: Century).

Elliott, L. *et al.* (1983), *Industrial Hygiene Characterization of Commercial Applications of Genetic Engineering and Biotechnology* (Washington, DC: NIOSH).

Ellul, J. (1965), *The Technological Society* (London: Cape).

Eysenck, H. (1971), *Race, Intelligence and Education* (London: Temple Smith).

Feder, J. and Tolbert, W.R. (1983), 'The large-scale cultivation of mammalian cells', *Scientific American* 248(1): 24–31.

Fedoroff, N.V. (1984), 'Transposable genetic elements in maize', *Scientific American* 250 (6): 65–75.

Fishlock, D. (1982), *The Business of Biotechnology* (London: Financial Times Business Information).

Fleming, D. (1968), 'Emigré physicists and the biological revolution,' *Perspectives in American History* 152–89.

Frank, P. (1949), *Modern Science and Its Philosophy* (Cambridge, Mass.: Harvard University Press).

Freifelder, D. (ed.) (1978), *Recombinant DNA: Readings from Scientific American* (San Francisco: Freeman).

Gallo, R.C. *et al.* (1984), 'Serological analysis of a subgroup of human T-lymphotropic retroviruses (HTLV–III) associated with AIDS', *Science* 224: 503–5.

Galton, F. (1869), *Hereditary Genius: An Inquiry into Its Laws and Consequences* (London: Macmillan).

Galton, F. (1883) *Inquiries into Human Faculty and Its Development* (London: Macmillan).

Garrod, A.E. (1963), *Inborn Errors of Metabolism*, reprinted with a supplement by H. Harris (London: Oxford University Press).

Gershon, S. (1983), 'Should science be stopped? The case of recombinant DNA research', *NIMH, Public Interest* 71: 3–16.

Gibbons, M. and Wittrock, B. (eds.) (1985), *Science as a Commodity* (Harlow: Longman).

GMAG (1978), 'Genetic manipulation: new guidelines for UK', *Nature* 276: 104–8.

Godber Report (1975), *Report of the Working Party on the Laboratory Use of Dangerous Pathogens* (London: HMSO, Cmnd 6054).

Goldberg, T. (1985), 'Moving toward public participation in biotech', *GeneWATCH* 2(3): 1–10.

Gonick, L. and Wheelis, M. (1984). *The Cartoon Guide to Genetics* (London: Harper & Row).

Gould, S.J. (1980), *The Panda's Thumb* (New York: Norton).

Gould, S.J. (1981), *The Mismeasure of Man* (New York: Norton).

Gribbin, J. (1985), *In Search of the Double Helix* (London: Corgi).

Grobstein, C. (1977), 'The recombinant-DNA debate', *Scientific American* 237(1): 22–33.

Grobstein, C. (1979), *A Double Image of the Double Helix* (San Francisco: Freeman).

Grobstein, C. *et al.* (1983), 'External human fertilization: An evaluation of policy', *Science* 222: 127–33.

Guidelines for Research Involving Recombinant DNA Molecules (1976) (Washington, DC: US Department of Health, Education, and Welfare, National Institutes of Health.)

Guidelines for Research Involving Recombinant DNA Molecules (1978) (Washington, DC: US Department of Health, Education, and Welfare, National Institutes of Health).

Guidelines for Research Involving Recombinant DNA Molecules (Guidelines or Rules) (1980) (Washington, DC: US Department of Health and Human Services, National Institutes of Health).

Guidelines for Research Involving Recombinant DNA Molecules (1982) (Washington, DC: Department of Health, Education, and Welfare, National Institutes of Health).

Gurdon, J.B. (1977), 'Egg cytoplasm and gene control', *Proceedings of the Royal Society London B* 198: 211–47.

Guthrie, R. (1973), 'Mass screening for genetic disease', in V.A. McKusick and R. Claiborne (eds.), *Medical Genetics* (New York: HP Publishing), pp. 229–37.

Haberer, J. (1972), 'Politicalization in science', *Science* 178: 713–24.

Hadlington, S. (1987), 'Transgenic sow', *Nature* 328: 6.

Hall, B.G. *et al.* (1983), 'Role of cryptic genes in microbial evolution', *Molecular Biology Evolution* 1: 109–24.

Halvorson, D. *et al.* (eds.) (1985), *Engineered Organisms in the Environment: Scientific Issues* (Washington, DC: American Society for Microbiology).

Hammer, R.E. *et al.* (1984), 'Partial correction of murine hereditary growth disorder by germ-line incorporation of a new gene', *Nature* 311: 65–7.

Harsanyi, Z. and Hutton, R. (1982), *Genetic Prophecy: Beyond the Double Helix* (London: Granada).

Harris, R. and Paxman, J. (1982), *A Higher Form of Killing* (London: Chatto & Windus).

Hayward, W.S. *et al.* (1981), 'Activation of cellular *onc* gene by promoter insertion in ALV-induced lymphoid leukosis', *Nature* 290: 475–80.

Health and Safety Commission (1987), *Review of the Health and Safety (Genetic Manipulation) Regulations 1978* (London: HSE).

Heden, C.G. (1982), 'Information on microbiology', *Impact of Science on Society* 32(2): 133–40.

Henifin, M.S. and Hubbard, R. (1983), 'Genetic screening in the workplace', *GeneWATCH* 1(1): 5–9.

Herbert, F. (1982), *The White Plague* (New York: GP Putnam's Sons).

Herskowitz, I.H. (1977), *Principles of Genetics* (New York: Macmillan).

Hilder, V.A. *et al.* (1987), 'A novel mechanism of insect resistance engineered into tobacco', *Nature* 300: 160–3.

Howell, R.R. (1973), 'Genetic disease: the present status of treatment', in V.A. McKusick and R. Claiborne (eds.), *Medical Genetics* (New York: HP Publishing), pp. 271–82.

Hsia, D.Y. and Holtzman, N.A. (1973), 'A critical evaluation of PKU screening', in V.A. McKusick and R. Claiborne (eds.), *Medical Genetics* (New York: HP Publishing), pp. 237–45.

Human Fertilisation and Embryology: A Framework for Legislation (London: HMSO, Cmnd 259).

Illmensee, K. and Hoppe, P.C. (1981), 'Nuclear transplantation in Mus musculus: development potential of nuclei from preimplantation embryos', *Cell* 23: 9–18.

Jacobs, P.A. *et al.* (1967), 'Aggressive behaviour, mental subnormality and the XYY male', *Nature* 213: 815.

Jayaraman, K.S. (1987), 'Vaccines developed in United States to be tested in India', *Nature* 328: 287.

Jeffreys, A.J. *et al.* (1985a), 'Individual-specific "fingerprints" of human DNA', *Nature* 316: 76–9.

Jeffreys, A.J. *et al.* (1985b), 'Positive identification of an immigration test-case using human DNA fingerprints', *Nature* 317: 818–9.

Jensen, A.R. (1969), 'How much can we boost IQ and scholastic achievement?', *Harvard Educational Review* 39: 1–123.

Jewkes, J. (1972), 'Government and high technology', *Occasional Paper 37* (London: IEA).

Joyce, C. (1984a), 'Will the USSR take genes to war?', *New Scientist*, 31 May, p. 9.

Joyce, C. (1984b), 'US backs gene swaps in animals', *New Scientist*, 1 November, p. 7.

Joyce, C. (1986), 'US exports genetic experiment', *New Scientist*, 20 November, p. 15.

Joyce, C. (1987), 'The race to map the human genome', *New Scientist*, 5 March, pp. 35–9.

Judson, H.F. (1979), *The Eighth Day of Creation* (London: Cape).

Kaback, M.M. and O'Brien, J.S. (1973), 'Tay-Sachs: prototype for prevention of genetic disease', in V.A. McKusick and R. Claiborne (eds.), *Medical Genetics* (New York: HP Publishing), pp. 253–62.

Kallmann, F.J. (1938), 'Eugenic birth control in schizophrenic families', *Journal of Contraception* 3: 195–9.

Keller, E.F. (1983), *A Feeling for the Organism* (New York: W.H. Freeman & Co.).

Kingman, S. (1987), 'The quest for an AIDS vaccine', *New Scientist*, 27 August, pp. 24–5.

Kirkwood, E.M. and Lewis, C.J. (1983), *Understanding Medical Immunology* (Chichester: John Wiley & Sons).

Korwek, E. (1980), 'The NIH Guidelines for recombinant DNA research and the authority of FDA to require compliance with the guidelines', *Food Drug Cosmetic Law Journal* 35: 633–50.

Korwek, E. (1981), 'Recombinant DNA and the law: review of some general considerations', *GENE (AMST)* 15(1): 1–5.

Krimsky, S. (1982), *Genetic Alchemy: The Social History of the Recombinant DNA Controversy* (Cambridge, Mass.: MIT Press).

Krimsky, S. (1985), 'The corporate capture of genetic technologies', *Science for the People* 17(3): 32–7.

Kuhn, T.S. (1970), *The Structure of Scientific Revolutions* (Chicago: University of Chicago Press).

Lacy, E. *et al.* (1983), 'A foreign *B*-globin gene in transgenic mice: integration at abnormal chromosomal positions and expression in inappropriate tissues', *Cell* 34: 343–58.

Land, H. *et al.* (1983), 'Cellular oncongenes and multistep carcinogenesis', *Science* 222: 771–8.

Landrigan, P.J. *et al.* (1982), 'Medical surveillance of biotechnology workers: report of the CDC/NIOSH Ad Hoc Working Group on Medical Surveillance for Industrial Applications of Biotechnology', *Recombinant DNA Technical Bulletin* 5(3): 133–8.

Lappe, M. (1985), *Broken Code* (San Francisco: Sierra Club Books).

Laux, D.C. *et al.* (1982), 'The effect of plasmid gene expression on the colonizing ability of *E. coli* HS in mice', *Recombinant*

DNA Technical Bulletin 5: 1–5.

Leder, P. *et al.* (1983), 'Translocations among antibody genes in human cancer', *Science* 222: 765–71.

Levy, S.B. *et al.* (1980), 'Survival of *Escherichia coli* host-vector systems in the mammalian intestine', *Science 209: 391–4.*

Levine, M.M. *et al.* (1983), 'Recombinant DNA risk assessment studies in man: efficacy of poorly mobilizable plasmids in biological containment', *Recombinant DNA Technical Bulletin* 6: 89–97.

Lewontin, R.C. (1984), 'The structure of evolutionary genetics', in E. Sober (ed.), *Conceptual Issues in Evolutionary Biology* (Cambridge, Mass., and London: MIT Press), pp. 3–13.

Lichtenstein, C. (1987), 'Bacteria conjugate with plants', *Nature* 328: 108–9.

Lindegren, C.C. (1966), *The Cold War in Biology* (Ann Arbor: Planarian Press).

Mayer, M. (1987), 'Genetic engineering and the environment: an update', *ENDS Report* 148: 9–12.

McClintock, B. (1978), 'Mechanisms that rapidly reorganise the genome', *Stadler Genetics Symposium* 10: 25–48.

McClintock, B. (1984), 'The significance of responses of the genome to challenge', *Science* 226: 792–801.

McCormick, D. (1985), 'Tuning the advisory mechanism', *Bio/Technology* 3 (April): 279.

McGrath, J.P. *et al.* (1984), 'Comparative biochemical properties of normal and activated human *ras* p21 protein', *Nature* 310: 644–9.

Mackenzie, D. (1984), 'Life's coming of age in utero', *New Scientist*, 9 August, p. 42.

McKusick, V.A. (1969), *Human Genetics* (Englewood Cliffs, NJ: Prentice-Hall).

McNally, R.M. and Wheale, P.R. (1986), 'Recombinant DNA technology: re-assessing the risks', *STSA Newsletter*, Summer, pp. 56–69.

Magner, L.N. (1979), *A History of the Life Sciences* (New York and Basel: Marcel Dekker).

Malthus, T.R. (1798), *An Essay on the Principle of Population as It Affects the Future Improvement of Society with Remarks on the Speculations of Mr. Godwin, M. Condorcet, and Other Writers*, 1st edn (London; reprinted, London: Macmillan, 1909).

Maniatis, A. *et al.* (1982), *Molecular Cloning: A Laboratory Manual* (New York: Cold Spring Harbor Laboratory).

Manning, D.H. (1977), *Society and Food* (London: Butterworths).

Marx, J.L. (1981), 'Three mice "cloned" in Switzerland', *Science* 211: 375–6.

Marx, J.L. (1983), 'Human T-cell leucaemia virus linked to AIDS', *Science* 220: 806–9.

Marx, J.L. (1984), 'Strong new candidate for AIDS agent', *Science* 224: 475–7.

Marx, J.L. (1986), 'The slow, insidious natures of the HTLVs', *Science* 231: 450.

Massey, R.J. (1987), 'Catalytic antibodies catching on', *Nature* 328: 457–8.

Medawar, P.B. and Medawar, J.S. (1977), *The Life Science* (London: Granada).

Merritt, G. (1987). 'Unnatural selection', *Weekend Financial Times*, 5 September, p. 1.

Milewski, E. (1985), 'Discussion on a proposal to form a RAC working group on biological weapons', *Recombinant DNA Technical Bulletin* 8(4): 173–5.

Military Construction, Army (1984), Reprogramming Request, Project 0817200, p. 1.

Mill, J.S. (1859), *Essay On Liberty* (London).

Mill, J.S. (1885), *Three Essays on Religion* (London: Longmans, Green & Co.).

Miller, J. and Loon, B. Van (1982), *Darwin for Beginners* (London: Writers and Readers Publishing Co-operative Society).

Minden, S. (1985), 'Genetic engineering and human embryos', *Science for the People* 17(3): 27–31.

Monod, J. (1971), *Chance and Necessity* (Glasgow: Collins).

Montagnier, L. *et al.* (1983), 'Isolation of a T-lymphotropic retrovirus from a patient at risk for Acquired immune deficiency syndrome (AIDS)', *Science* 220: 868–71.

Mooney, H.A. and Drake, J.A. (eds.) (1986), *Ecology of Biological Invasions of North America and Hawaii* (New York: Springer-Verlag).

Morrow, J.F. *et al.* (1974), 'Replication and transcription of eucaryotic DNA in *Escherichia coli*', *Proceedings of the National Academy of Sciences* 71: 1743–7.

Motulsky, A.G. (1984), 'Impact of genetic manipulation on society and medicine', *Science* 219: 135–40.

Muller, H.J. (1935), *Out of the Night: A Biologist's View of the Future* (New York: Vanguard Press).

Muller, H.J. (1963), 'Genetic progress by voluntarily conducted germinal choice', in G. Wolstenholme (ed.), *Man and His Future* (London: J. & A. Churchill), pp. 247–62.

Murphy, S. *et al.* (1984), *No Fire, No Thunder* (London: Pluto Press).

Murray, T.H. (1985), 'Ethical issues in genetic engineering', *Social Research* 52(3): 471–89.

Nasmyth, K. and Sulston, J. (1987), 'High-altitude walking with YACs', *Nature* 328: 380–1.

Newman, S. (1985), 'What are genes?', *GeneWATCH* 2(1): 12–13.

Newmark, P. (1987a), 'UK releases of engineered organisms to go ahead', *Nature* 327: 265.

Newmark, P. (1987b), 'Britain's genetic manipulation regulations to be extended', *Nature* 329: 379.

NIH, ORDA (1983), Minutes of RAC meeting of 11 April, p. 15.

NIH Working Group on Revision of the Recombinant DNA Guidelines, (1981), 'Evaluation of the risks associated with recombinant DNA research', *Federal Register*, 4 December (46: 59385–94).

Nielsen, J. (1968), 'The XYY syndrome in a mental hospital: genetically determined criminality', *British Journal of Criminality* 8: 186.

Oka, I.N. and Pimentel, D. (1976), 'Herbicide (2,4–D) increases insect and pathogen pests on corn', *Science* 193: 239–40.

Olby, R. (1974), *The Path to the Double Helix* (London: Macmillan).

Olejnikowna, I. (ed.) (1986), 'Biotechnology resources for schools and colleges', *ASE Occasional Papers Series* (Hatfield, Herts: Association for Science Education).

ORDA (Office of Recombinant DNA Activities) (1986), 'Points to consider in the design and submission of human somatic-cell gene therapy protocols', *Recombinant DNA Technical Bulletin* 9(4): 221–42.

OTA (1984a), *Commercial Biotechnology: An International Analysis* (Washington, DC: Office of Technology Assessment).

OTA (1984b), *Human Gene Therapy* (Washington DC: Office of Technology Assessment Background Paper, GPO stock number 052–003–00983–8)

Pain, R.H. (1987), 'A case of mistaken identity', *Nature* 328: 298.

Palca, J. (1986), 'Viral cancers: AIDS virus at the centre', *Nature* 319: 170.

Palca, J. (1987a), 'AIDS virus infects another lab worker', *Nature* 329: 571.

Palca, J. (1987b), 'Changing features sighted on the biotechnology horizon', *Nature* 330: 512.

Patent Profiles: Biotechnology (1982) (Washington, DC: Office of Technology Assessment and Forecast for the US PTO).

Pettinan, R. (1980), *Biopolitics and International Values* (Oxford: Pergamon Press).

Piller, C. (1985), 'Biological warfare and the new genetic technologies', *Science for the People* 17(3): 10–15.

Prentis, S. (1984), *Biotechnology: A New Industrial Revolution* (London: Orbis).

Prigogine, I. and Stengers, I. (1984), *Order Out of Chaos* (London: Heinemann).

Reilly, P. (1977), *Genetics, Law and Social Policy* (Cambridge, Mass.: Harvard University Press).

Rifkin, J. (1984), *Algeny: A New World* (London: Pelican).

Robertson, M. (1986), 'Agricultural genetics still on ice', *Nature* 320: 571.

Robinson, J.P. (1982), 'The changing status of chemical and biological warfare: recent technical, military and political developments', in *SIPRI Yearbook* (London: Taylor Francis), p. 324.

Roitt, I.M. *et al.* (1985), *Immunology* (Edinburgh: Churchill Livingstone).

Rose, M.R. and Doolittle W.F. (1983), 'Molecular biological mechanisms of speciation', *Science* 220: 157–62.

Rose, S. (1976), 'Scientific racism and ideology: the IQ racket from Galton to Jensen', in H. Rose and S. Rose, *The Political Economy of Science* (London: Macmillan).

Rose, S. (1981), 'The case against chemical weapons', *New Scientist*, 12 March, pp. 670–1.

Rose, S. *et al.* (1984), *Not in our Genes* (Harmondsworth: Penguin).

Rose, S. (1987), 'Biotechnology at war', *New Scientist*, 19 March, pp. 33–7.

Rosenberg, G. and Simon, L. (1979), 'Recombinant DNA: have recent experiments assessed all the risks?' *Nature* 282: 773–4.

Rosenblatt, D.P. (1982), 'The regulation of recombinant DNA research: the alternative of local control', *Boston College Environment Affairs Law Review* 10(1): 37–78.

Rothman, H. *et al.* (1980), *Biotechnology: A Review and Annotated Bibliography* (London: Frances Pinter).

Rothwell, R. and Zegveld, W. (1985), *Reindustrialisation and Technology* (Harlow: Longman).

Rowley, C.K. (1974), 'Pollution and public theory', in A.J. Culyer (ed.), *Economic and Social Goals* (London: Martin Robertson), pp. 284–312.

Royal Society Report (1981), *Biotechnology and Education: The Report of a Working Group* (London: The Royal Society).

Sagik, B.P. *et al.* (1981), 'The survival of EK1 and EK2 systems in sewage treatment plant models', in S.B. Levy *et al.* (eds.), *Molecular Biology, Pathogenicity and Ecology of Bacterial Plasmids* (New York: Plenum Press), pp. 499–60.

Saliwanchik, R. (1982), *Legal Protection for Microbial and Genetic Engineering Inventions* (Reading, Mass.; London: Addison-Wesley).

Salomon, J.J. (1973), *Science and Politics* (London: Macmillan).

Sayre, A. (1978), *Rosalind Franklin & DNA* (New York: Norton).

Schneider, K. (1986a), 'Release of a gene-altered agent violated policy, EPA charges', *The New York Times*, 26 February, p. B28.

Schneider, K. (1986b), 'Questions raised over how US backed genetically altered virus', *The New York Times*, 4 April, pp. A1, A20.

Schumpeter, J.A. (1939), *Business Cycles: A Theoretical, Historical and Statistical Analysis of the Capitalist Process* (New York: McGraw-Hill).

Schroedinger, E. (1945), *What is Life?* (Cambridge: Cambridge University Press).

Scott, A. (1986a), 'Viruses and cells: a history of give and take?', *New Scientist*, 16 January, pp. 42–5.

Scott, A. (1986b), 'At the limits of infection,' *New Scientist*, 15 May, pp. 41–4.

Scott, A. (1987), *Pirates of the Cell: The Story of Viruses from Molecule to Microbe*, 2nd edn (Oxford: Basil Blackwell).

Seagrave, S. (1982), *Yellow Rain* (London: Sphere).

Shapiro, J. (ed.) (1983), *Mobile Genetic Elements* (New York: Cold Spring Harbor Symposium).

Sharples, F.E. (1987), 'Regulation of products from biotechnology', *Science* 235: 1329–32.

Simons, J.P. *et al.* (1987), 'Alteration of the quality of milk by expression of sheep beta-lactoglobulin in transgenic mice', *Nature* 328: 530–2.

Singer, M.F. (1983), 'The genetic program of complex organisms', in *Frontiers in Science and Technology: A Report by the Committee on Science, Engineering and Public Policy of the National Academy of Sciences, National Academy of Engineering and Institute of Medicine* (New York: Freeman), pp. 15–44.

Singer, M.F. and Soll, D. (1973), 'Guidelines for DNA hybrid molecules', *Science* 181: 1114.

Singer, P. and Wells, D. (1984), *The Reproduction Revolution* (Oxford: Oxford University Press).

Sinsheimer, R. (1979), 'Two lectures on recombinant DNA research', in D.A. Jackson and S.P. Stick (eds.), *The Recombinant DNA Debate* (Englewood Cliffs, NJ: Prentice-Hall).

SIPRI (Stockholm International Peace Research Institute) (1973), *The Problem of Chemical and Biological Warfare, Vol. V: The Prevention of Chemical and Biological Warfare* (New York: Humanities Press).

Smith, J.E. (1981), *Biotechnology* (London: Edward Arnold).

Smith, R.J. (1984a), 'The dark side of biotechnology', *Science* 224: 1215–16.

Smith, R.J. (1984b), 'New army biowarfare lab raises concerns', *Science* 226: 1176–8.

Sober, E. (ed.) (1984), *Conceptual Issues in Evolutionary Biology* (Cambridge, Mass., and London: MIT Press).

Sonea, S. and Panisset, M. (1980), *A New Bacteriology* (Boston, Mass.: Jones & Bartlett).

Southwick, C.H. (1976), *Ecology and the Quality of Our Environment* (New York: D. Van Nostrand Company).

Spier, R.E. (1982), 'Animal cell technology: an overview', *Journal of Chem. Tech. Biotechnol.* 23: 304–12.

Stacey, D.W. and King, H.F. (1984), 'Transformation of NIH 3T3 cells by microinjection of Ha-*ras* p21 protein', *Nature* 310: 508–11.

Stackbrandt, E. and Woese, C.R. (1984), 'The phylogeny of prokaryotes', *Microbiological Sciences* 1: 117–22.

Stent, G.S. (1971), *Molecular Genetics: An Introductory Narrative* (San Francisco: Freeman).

Strauss, H. *et al.* (1986), 'Genetically-engineered micro-organisms: I. Identification, Classification and Strain History', *Recombinant DNA Technical Bulletin* 9(1): 1–15.

Sun, M. (1984), 'Committee vetoes proposal to ban gene tests', *Science* 226: 674.

Sun, M. (1985), 'Monsanto may bypass NIH in microbe test', *Science* 227: 153.

Sylvester, E.J. and Klotz, L.C. (1983), *The Genetic Age: Genetic Engineering and the New Industrial Revolution* (New York: Charles Scribner's Sons), Ch.6, pp. 115–26.

Szybalski, W. (1985), 'Genetic engineering in agriculture, Letters', *Science* 229: 112–15.

Timmis, K.N. and Puhler, A. (eds.) (1979), *Plasmids of Medical, Environmental and Commercial Importance* (Amsterdam: Elsevier/North-Holland Biomedical Press).

Tonegawa, S. *et al.* (1978), 'Organisation of immunoglobulin genes', *Cold Spring Harbor Symposium of Quantitative Biology* 42: 921–31.

United Kingdom, Health and Safety *(Genetic Manipulation) Regulations* (1978) (London: HMSO).

UPOV (1981), *Geneva Diplomatic Conference in the Revision of the International Convention for the Protection of New Varieties of Plants (1978)* (Geneva: UPOV).

Vaeck, M. *et al.* (1987), 'Transgenic plants protected from insect attack', *Nature* 328: 33–7.

Vines, G. (1986), 'New tools to treat genetic diseases', *New Scientist*, 13 March, pp. 40–2.

Vogel, S. (1985), 'Biogenetic waste: a federal and local problem', *GeneWATCH* 2 (4–6): 7–9.

Wade, N. (1975), 'Genetics: conference sets strict rules to replace moratorium', *Science* 187: 931–15.

Wade, N. (1980a), 'University and drug firm battle over billion-dollar gene', *Science* 209: 1492–4.

Wade, N. (1980b), 'UCSD gene splicing incident ends unresolved', *Science* 209: 1494–5.

Wade, N. (1981), 'Gene therapy caught in more entanglements', *Science* 212: 24–5.

Walton, A.G. and Hammer, S. (1983), 'Stanford, Cohen-Boyer patent', *Genetic Engineering and Biotechnology Yearbook* (New York: Elsevier Science Publishers BV).

Waneck, G. (1985), 'Safety and health issues revisited', *Science for the People* 17(3): 38–43.

Watson, J.D. (1968), *The Double Helix* (New York: Atheneum).

Watson, J.D. (1987), 'Minds that live for science', *New Scientist*, 21 May, pp. 63–6.

Watson, J.D. and Crick, F.H.C. (1953), 'Molecular structure of nucleic acids: a structure for deoxyribose nucleic acid', *Nature* 171: 737–8.

Watson, J.D. and Tooze, J. (1981), *The DNA Story: A Documentary History of Gene Cloning* (New York: Freeman).

Watson, J.D. *et al.* (1983), *Recombinant DNA* (New York and Oxford: Freeman).

Weatherall, D.J. (1985), *The New Genetics and Clinical Practice* (Oxford: Oxford University Press).

Wells, H.G. (1905), *A Modern Utopia* (London: Chapman & Hall).

Wheale, P.R. (1987), 'Test tube bugs on the loose', *Daily Mail*, 20 May, p. 6.

Wheale, P.R. and McNally, R.M. (1984), *Rat and Super Rat* (unpublished play).

Wheale, P.R. and McNally, R.M. (1986), 'Patent trend analysis: the case of microgenetic engineering', *Futures*, October, pp. 638–57.

Williams, D.A. *et al.* (1984), 'Introduction of new genetic material into pluripotent haemopoietic stem cells of the mouse', *Nature* 310: 476–80.

Williams, J.G. (1982), 'Mouse and supermouse', *Nature* 300: 575.

Williams Report (1976), *Report of the Working Party on the Practice of Genetic Manipulation* (London: HMSO, Cmnd 6600).

Wootton, B. (1963), *Crime and the Criminal Law: Reflections of a Magistrate and Social Scientist* (London: Stevens).

WHO (1970), *Health Aspects of Chemical and Biological Weapons* (Geneva: WHO).

WHO (1987), *An International Study on the Occurrence of Multi-resistant Bacteria and Aminoglycoside Consumption Patterns* (Copenhagen: WHO, ATH).

Wright, K. (1986), 'New regulations in dispute', *Nature* 323: 387.

Wright, P. (1987), 'New peril looms as routine vaccination activates Aids virus', *The Times*, 20 April, p. 3.

Wright, S. (1960) in National Academy of Sciences, The Biologist Effects of Atomic Radiation (Washington DC: National Research Council).

Wright, S. (1985), 'The military and the new biology', *Bulletin of the Atomic Scientists* 41(5): 10–16.

Wright, S. (1986), 'Molecular biology or molecular politics? The production of scientific consensus on the hazards of recombinant DNA technology', *Social Studies of Science* 16: 593–620.

Yanchinski, S. (1985), *Setting Genes to Work: The Industrial Era of Biotechnology* (London: Viking).

Yoxen, E. (1983), *The Gene Business* (London: Pan Books).

Yoxen, E. (1986), *Unnatural Selection: Coming to Terms with the New Genetics* (London: Heinemann).

Yoxen, E. (1987), 'Report of a Four-Nation Study of the Social and Environmental Impact of Biotechnology', *European Foundation*.

Yunis, J.J. (1983), 'The chromosomal basis of human neoplasia', *Science* 221: 227–36.

Glossary

Amino acids: The building blocks of protein structure; twenty different amino acids are commonly found in living organisms.

Antibiotic: (Gr. *ante*, against or opposite; *bios*, life.) A substance capable of killing or preventing the growth of a micro-organism; can be produced by another micro-organism or synthetically.

Antibody: A protein produced in the body in response to the presence of a foreign chemical substance or organism (antigen), shaped to fit precisely to the antigen and in such a way as to annul its action or help to destroy it; part of the body's defence (immune) system.

Antigen: A substance which causes the immune system of the body to manufacture specific antibodies that will react with it.

Artificial insemination (AI): The application of sperm to an unfertilized egg by means other than ejaculation during sexual intercourse.

Autosome: A chromosome other than a sex chromosome.

Bacteria: (Gr. *bacterium*, a little stick.) The simplest organisms that can reproduce unaided; a class of single-celled micro-organism found living freely in water, the soil and in the air, and as parasites within plants and animals; *Escherichia coli (E. coli)* is the species most commonly used as a host cell in recombinant DNA work.

Bacteriophage: (Gr. *phagein*, eat.) Viruses which infect bacteria; they are commonly referred to as a phage; bacteriophage are used as vectors in recombinant DNA experiments.

Base: Part of the building blocks of nucleic acids, the sequence of which encodes genetic information; cytosine (C), guanine (G), adenine (A) and thymine (T) are the bases in DNA; C. G, A and uracil (U) are the bases in RNA (see DNA BASE PAIR).

Biochemistry: The study of the chemistry of living things.

Biogenetic waste: Biological waste that contains genetically modified organisms; includes sewage, refuse and effluent from biotechnological processes.

Biological containment: The use of genetically-deficient microbial host cells, which are unlikely to survive outside the laboratory, and genetically-deficient cloning vectors, which are deficient in their ability to move to a new host strain.

Biosensor: The combination of a biological component sensitive to specific changes (e.g. in a bioreactor) and an electrical component which is able to translate the response of the biological component into a measureable signal.

Biosphere: (Gr. *bios*, life; *sphaira*, a ball or a globe.) All life that is encompassed under the vault of the sky; the part of the earth that is inhabited by living organisms; the Earth's surface and the top layer of the hydrosphere (water layer) have the greatest density of living organisms.

Biotechnology: (After Bull *et al*. 1982.) The application of scientific and engineering principles to the processing of materials by biological agents to provide goods and services.

Caducean: (L. *caduceus*, a winged rod entwisted with two serpents in the form of a double helix, and carried by Mercury, the swift ready messenger of the gods, and god of eloquence, theft, merchandise and trade.) Belonging to Mercury's wand.

Catalyst: A substance which assists a chemical reaction without itself being used up in the process.

Cell: (L. *cella*, a little room, from *cello*, hide.) In 1655 Robert Hooke (1625–1702), curator of the Royal Society, used the term 'cell' to describe the small, closed cavities he found upon microscopic examination of the outer bark of an oak tree; although the structures he observed were actually cell walls, which are absent from animal cells, the term has persisted to describe the basic unit of structure of all living organisms excluding viruses; as defined by Max Schultze (1825–74), a cell is 'a lump of nucleated protoplasm'; a generalized description of a cell is a mass of jelly-like cytoplasm, contained within a semi-permeable membrane, and containing a spherical body called the nucleus; a single cell constitutes the entire organism of a single-celled creature such as a bacterium; a human being is composed of millions of cells.

Cell fusion: The fusing together of two or more cells to produce a single hybrid cell.

Chimaera: L. *chimaera*, a mythological monster with the head of a lion, the body of a goat, and the tail of a dragon, vomiting flames.) An organism, cell or molecule (e.g. DNA) constructed from material from two different individuals or species.

Chromosomes: (Gr. *chroma*, colour; *soma*, body.) Darkly staining structures, composed of DNA and protein, which bear and transmit genetic information; they are found in the nucleus of eucaryotic cells and free in the cytoplasm in the cells of procaryotes; the number of chromosomes is constant for the somatic cells of a species; in human beings the normal chromosomal constitution is twenty-two pairs of autosomal chromosomes, and one pair of sex chromosomes.

Clone: A collection of genetically identical molecules, cells or organisms which has been derived asexually from a single common ancestor.

Cloning: Making identical copies of proteins, RNAs, genes, nuclei, cells and individuals by an asexual, biological process.

Codon: Three successive nucleotides (or bases) which specify a particular amino acid or a 'punctuation mark' in the genetic code.

Colony (of micro-organisms): A dense mass of micro-organisms produced asexually from an individual micro-organism.

Congenital disorder: A malfunction which is present at birth; the term describes all deformities and other conditions that are present at birth whether they are inherited or newly arisen as a result of adverse environmental factors.

Crucible: A small earthen pot, used by chemists, founders and others for melting purposes; a hollow place at the bottom of a furnace to receive the melted metal; a situation which severely tests a person's virtue.

Cytogenetics: The area of study that links the structure and behaviour of chromosomes with inheritance.

Cytoplasm: (Gr. *kytos*, hollow; *plasma*, mould.) The living contents of a cell excluding the nucleus.

Diffusion (of an innovation): The spread of an innovation, with or without modification, through a population of potential users.

Deoxyribo(se)nucleic acid (DNA): The molecule which for all organisms except RNA viruses encodes information for the reproduction and functioning of cells, and for the replication of the DNA molecule itself; DNA molecules are able to encode genetic information for the control system of the cell or organism which is transmitted from generation to generation.

DNA base pair: A pair of DNA nucleotide bases; one of the pair is on one chain of the duplex DNA molecule, the other is on the complementary chain; they pair across the double helix in a very specific way: adenine (A) can only pair with thymine (T); cytosine (C) can only pair with guanine (G); the specific nature of base-pairing enables accurate replication of the chromosomes and thus maintains the constant composition of the genetic material.

DNA probe: A short piece of DNA that is used to detect the presence of a specific sequence of bases on a DNA molecule, used in genetic screening for example.

DNA sequence: The order of base pairs in the DNA molecule; genetic information can be encoded in the DNA sequence.

DNA technology: See MICROGENETIC ENGINEERING.

Dominant genetic disorder: (L. *dominans*, ruling.) A disorder which is expressed in the phenotype when the gene responsible is inherited from just one parent (cf. RECESSIVE GENETIC DISORDER).

Double helix: The name given to the structure of the DNA molecule, which is composed of two complementary strands which lie alongside and twine around each other, joined by cross-linkages between base pairs (see DNA BASE PAIR).

Downstream process: A process in industrial biotechnology which occurs after the bioconversion stage; for example product recovery, separation and purification.

Duplicative transposition: The process whereby transposable genetic elements move around genomes; a copy of a transposable genetic element located on a chromosome is duplicated and then deposited at a new location without loss of the original sequence.

Ecosystem: (Gr. *oikos*, home; L. *systema*, an assemblage of things adjusted into a regular whole.) A unit made up of all the living and non-living components of a particular area that interact and exchange materials with each other.

Enablement requirement: A patentor's legal obligation to provide full technical details of his or her novel process or product, which should allow a person with ordinary skill in the field to duplicate the invention.

Endotoxins: Toxic substances formed inside the cells of bacteria that are released on disintegration of the cell and can cause adverse effects, such as fevers, in their hosts.

Enteric: (Gr. *enteron*, intestine.) Of, or pertaining to, the intestine.

Enzyme: (Gr. *en*, in; *zyme*, leaven.) A biological catalyst produced by living cells; a protein molecule which mediates and promotes a chemical process without itself being altered or destroyed; enzymes act with a given compound, the substrate, to produce a complex, which then forms the products of the reaction; enzymes are extremely efficient catalysts and very specific to particular reactions; the active principle of a ferment.

Escherichia coli (E. coli): A bacterial species that inhabits the intestinal tract of most vertebrates and on which much genetic work has been done; some strains are pathogenic to humans and other animals; many non-pathogenic strains are used experimentally as hosts for recombinant DNA.

Eucaryotes: Cells or organisms whose DNA is organized into chromosomes with a protein coat and sequestered in a well-defined cell nucleus; all living organisms except viruses, bacteria and blue-green algae are eucaryotic (cf. PROCARYOTES).

Eugenics: (Gr. *eu*, well; *genos*, birth.) (After Galton 1883.) The science which deals with all the influences that improve the inborn qualities of a race, also with those that develop them to the utmost advantage.

Evolution (biological): (L. *evolvere*, to unroll.) Changes in DNA that occur during the history of organisms; the development of new organisms from pre-existing organisms since the beginning of life.

Expression (of genes): see GENE EXPRESSION.

Feedstock: The raw material used for the production of chemicals.

Fermentation: The anaerobic (without oxygen) biological conversion of organic molecules, usually carbohydrates, into alcohol, lactic acid and gases; it is brought about by enzymes which are used for industrial purposes either directly or as components of certain bacteria and in yeasts; in general use, the term is applied to the biological reaction in other biotechnological processes.

Fertilization: In sexually reproducing animals, the activation of the development of an egg through the union of sperm with the egg, so combining their genetic complements.

Gene: (Gr. *genos*, descent.) A gene is a section of a nucleic acid molecule in which the sequence of bases encodes the structure of, or is involved in the synthesis of, a protein.

Gene enhancement: The insertion of additional genetic material into an otherwise normal genome in order to enhance a trait perceived as desirable; an example is the insertion of additional growth hormone genes into the genome of a normal individual in order to increase his or her height.

Gene expression: The mechanism whereby the genetic instructions in a given cell are decoded and processed into the final functioning product, usually a protein.

Gene mapping: Determining the relative locations (loci) of genes on chromosomes.

Gene therapy: The correction of the effect of a genetic defect in an organism or cell by direct intervention with the genetic material; one method under investigation is gene replacement therapy in which additional foreign DNA is inserted to compensate for the malfunctioning gene; another approach would be to activate dormant genes within the genome whose function would substitute for the missing function of the malfunctioning gene.

Genetic code: The relationship between the sequence of bases in the nucleic acids of genes and the sequence of amino acids in the proteins that they code for.

Genetic disorder: A disorder which is associated with a specific defect in the hereditary material; may or may not be congenital (e.g. late-onset genetic disorders are not manifested at birth) and may or may not be inherited (e.g. chromosomal abnormalities which are newly-arisen).

Genetic engineering: The manipulation of heredity or the hereditary material; the direct and deliberate attempts by humans to influence the course of evolution and to alter its products; the basic techniques are mutagenesis and hybridization, which introduce genetic variation, and artificial selection which biases qualitatively the genetic variation of subsequent generations; genetic engineering using artificial selection and hybridization has been long practised; artificial mutagenesis is a twentieth-century development; in the second half of the twentieth-century all three techniques have been refined to give the genetic engineer more control and to increase the speed with which change in a given direction can be achieved (see also MICROGENETIC ENGINEERING).

Genetic manipulation: See GENETIC ENGINEERING.

Genetic recombination: The excision and rejoining of DNA molecules; formation of a new association of genes or DNA sequences from different parental origins. After Herskowitz 1977: change in sequence or grouping of genetic nucleotides; changes in sequence including making and breaking concate-mers, integration and excision, crossing over, and gene conversion; changes in grouping include the separation of sister chromosomes in procaryotic and eucaryotic cell divisions; segregation, and independent segregation in meiosis; and conjugation, cell fusion and fertilization (see also RECOMBI-NANT DNA TECHNIQUES, AND IN VITRO GENETIC RECOMBINATION).

Genetics: (L. *genesis*, origin, descent.) The study of the information system of the reproduction, growth and functioning of the cell, and the replication and transmission of hereditary information;

that part of biology dealing with both the constancy of inheritance and its variation.

Genome: A collective noun for all the genetic information that is typical of a particular organism; every cell in a multicellular organism contains a full genome; the term genome is also applied to the genetic contents characteristic of major groups (e.g. the eucaryotic genome) or of a species (e.g. the human genome); not all portions of a genome are genes (i.e. genomes include non-coding DNA); genomes do not include the genetic material of extrachromosomal elements, nor of plasmids or viruses harboured by a cell, although the distinction in the case of dormant viruses may be semantic.

Genotype: (Gr. *genos*, race; *typos*, image.) The genetic constitution of an organism or group of organisms; the actual appearance and metabolism of an individual (the phenotype depends on the dominance relationships between alleles in the genotype and the interaction between genotype and environment (cf. PHENOTYPE).

Germ: A popular word for a micro-organism; germ was a common and rather vague term in biology in the eighteenth and nineteenth centuries and was used to describe the material of heredity.

Germ cell: Cells which give rise to sex cells, which in animals are sperm and eggs.

Germ-line: Cells from which sex cells are derived.

Germ-plasm: The term for that part of an organism which passes on hereditary characteristics to the next generation. According to Weismann's germ-plasm theory of 1883, the germ-plasm is transmitted from generation to generation in germ cells.

Haemoglobin: (Gr. *haima*, blood; L. *globus*, a ball.) The red-coloured protein which binds and carries oxygen in red blood cells.

Hereditary disease: A disorder of the genetic material which is transmissible from generation to generation.

Hormone: A chemical messenger of the body carried in the bloodstream from the gland which secretes it to a target organ where it has a regulatory effect; an example is insulin.

Host: A cell (microbial, animal or plant) whose metabolism is used for the replication of a virus, plasmid or other form of foreign DNA, including recombinant DNA.

HTLV: Human T-cell leucaemia virus; later renamed human T-cell lymphotropic virus.

Hybridoma: A 'hybrid myeloma'; a cell produced by the fusion of an antibody-producing cell (lymphocyte) with a cancer cell (myeloma).

Hybridoma technology: Hybridoma technology is the technology of fusing antibody-producing cells with tumour cells to produce continuously proliferating cells which produce monoclonal antibodies.

Immobilized enzyme: An enzyme which is attached to an insoluble support system.

Immune system: The body's system of defence against invasion by foreign organisms and certain chemicals.

Inborn error of metabolism: A single-gene recessive disorder which results in the failure of a biochemical reaction in a metabolic pathway to occur.

Infection: The invasion of an organism, or part of an organism, by pathogenic viruses, bacteria or other micro-organisms.

Innovation: The first introduction of a new product, process or system into the ordinary commerical or social activity of a country.

Insulin: A hormone, the release of which lowers the level of glucose sugar in the blood.

Interferons: A class of proteins released by certain mammalian cells in response to various stimuli, including viral infection, which are thought to inhibit viral replication; undergoing clinical trials as anti-viral and anti-cancer agents.

Intron: A nucleic acid sequence within a gene, which is transcribed into RNA but then excised from the RNA transcript before it is translated into protein.

In utero: (L. *uterus*.) In the womb.

Invention: The first idea, sketch or contrivance of a new product, process or system.

In vitro: (L. *vitrum*, glass.) Literally in glass; biological processes studied outside the living organism.

***In vitro* fertilization (IVF)**: The fertilization of an egg cell by sperm on a glass dish (compared to the 'natural' *in vivo* fertilization of egg cells).

***In vitro* genetic recombination**: The precise excision and joining of DNA fragments on the laboratory bench exploiting the biochemical tools of the cell—restriction enzymes and DNA ligase—and the inherent pairing affinity of the duplex DNA molecule (see RECOMBINANT DNA TECHNIQUES and GENETIC RECOMBINATION).

In vivo: (L. *vivo*, live.) Within the living organism.

Karyotype: The chromosomal constitution of an individual.

Ligase: An enzyme which catalyses the joining together of two molecules; DNA ligase catalyses the joining of two DNA molecules.

Microbe: An alternative term for a micro-organism.

Microbiology: The study of micro-organisms.

Microgenetic engineering: The techniques which enable the genetic engineer to decode, compare, construct, mutate, excise, join, transfer and clone specific sequences of DNA, thus directly manipulating the genetic material to produce organisms, cells and subcellular components; applications include scientific research, biotechnology, farming, health care and biological defence.

Micro-organism: An organism belonging to the categories of viruses, bacteria, fungi, algae or protozoa; micro-organisms—'animalcules'—were first observed by Anton van Leeuwenhoek (1632—1723) of Delft in Holland in the seven-teenth-century; until the second half of the nineteenth-century, when the nature of their association with putrefaction, fermentation and disease became a focus of microscopy, Leeuwenhoek's discovery of parasites and bacteria remained a 'curiosity', and those who continued to search for *Vermes chaos*, as Linnaeus classified these 'incredibly small animals', were regarded as eccentrics.

Molecule: A group of two or more atoms joined together by chemical bonds in a specific way.

Mobile genetic elements: RNA and DNA sequences that move from one place to another both within and between genomes; the family of mobile genetic elements includes viruses, subviral infectious elements, plasmids, RNA introns, messenger RNA molecules, transposable genetic elements and oncogenes; characteristic features are the ability to move pieces of DNA either within or between cellular genomes, the ability to usurp the biochemistry of the host cell to bring about their own replication, the ability to alter the structure of the host cell genome, and the ability to modify the expression of other genetic elements within the host cell; they are believed to play a crucial role in evolution, and are implicated in some aspects of development and pathology; viruses and plasmids are mobile genetic elements which are used as vectors in gene transfer technology.

Monoclonal antibodies: A clone of one specific antibody produced by hybridomas.

Multifactorial disorder: A disorder in which both genetic and exogenous factors are multiple and interact; the genetic part of multifactoral causation is polygenic (the result of the action of many genes).

Mutagen: A chemical or physical agent which increases the frequency of mutation.

Mutant: An organism or gene which deviates by mutation from the parent organism(s) or gene in one or more characteristics.

Mutation: (L. *muto*, change.) A change in the genetic material; usually refers to changes in a single DNA base pair or in a single gene, but includes changes recognizable microscopically as chromosomal aberrations; in connection with inherited diseases and evolution, mutation in the germ-line or sex cells is most relevant, but somatic cell mutation also occurs and may be the basis of some cancers and some aspects of ageing.

New technological system: Where new industries have arisen; the diffusion of clusters of fundamental innovations conduce to the emergence of 'new technological systems'.

Nucleic acids: Either RNA or DNA; complex organic molecules composed of sequences of nucleotide bases.

Nucleus: A region in the cells of eucaryotes, surrounded by a membrane, in which the main chromosomes are sequestered.

Oncogene: (Gr. *onchos*, swelling; L. *genesis*, origin.) Found in every cell of the body, it is postulated that oncogenes are a broad class of regulatory genes that control the activity of other genes; the oncogene theory of cancer is that when an oncogene is activated at an innappropriate stage in the life-cycle of an individual, the cell begins to multiply in an uncontrolled way; cellular oncogenes (proto-oncogenes) may be derived from viruses which have integrated into the genome of a host cell, or conversely, certain viruses may have acquired cellular oncogenes when they became infectious.

Oncogenic: Cancer-causing.

Organic compound: A chemical compound which contains carbon atoms; originally discovered in living organisms.

Patent: The exclusive right to a property in an invention; this monopoly on invention gives its owner the legal right of action against anyone exploiting the patented research without the patentor's consent.

Pathogen: (Gr. *pathos*, suffering.) An organism which causes disease.

Phage: Abbreviation of *BACTERIOPHAGE*, a bacterial virus.

Phenotype: (Gr. *phainein*, to show; *typos*, image.) The collective physical properties of an organism determined by the interaction of its genetic constitution (genotype) and environmental experiences; many genes present in the genotype do not show their effects in the phenotype; genotypically identical organisms may have very different phenotypes in different environments (cf. GENOTYPE).

Physical containment: Measures that are designed to prevent or minimize the escape of recombinant micro-organisms or cells.

Plasmid: A small circle of DNA usually found in procaryotes, which replicates independently of the main chromosome(s) and can be transferred naturally from one organism to another, even across species boundaries; plasmids and some viruses are used as vectors for cloning recombinant DNA in bacterial host cells.

Polypeptide: Long folded chains of amino acids; proteins are made from polypeptides.

Procaryotes: Organisms whose genetic material is not sequestered in a well-defined nucleus; includes bacteria and blue-green algae (cf. EUCARYOTES).

Promoter: A DNA sequence to which the enzymes which catalyse messenger RNA synthesis bind.

Protein: Proteins are polypeptides, that is they are made up of amino acids joined together by peptide links; acting as hormones, enzymes and connective and contractile structures, proteins endow cells and organisms with their characteristic properties of shape, metabolic potential, colour and physical capacities.

Recessive genetic disorder: (L. *recessus*, withdrawn.) The gene mutations responsible for recessive disorders are usually only harmful to individuals who do not have a corresponding normal gene. Except in the case of sex-linked conditions, a recessive disorder is only expressed when the gene mutation responsible is inherited from both parents (see DOMINANT GENETIC DISORDER).

Recombinant bacterium/cell/plasmid/vector/virus (etc.): Contains recombinant DNA (or recombinant RNA).

Recombinant DNA: A hybrid DNA molecule which contains DNA from two distinct sources (see GENETIC RECOMBINATION).

Recombinant DNA techniques (technology): The combination of *in vitro* genetic recombination techniques with techniques for the insertion, replication and expression of recombinant DNA inside living cells.

Restriction enzymes: Bacterial enzymes that cut DNA at specific DNA sequences; exploited by the microgenetic engineer, for example, in the execution of the precise excision of DNA fragments for *in vitro* genetic recombination, and in the analysis of the base sequence of DNA molecules.

Retroviruses: Viruses which encode their hereditary information in the nucleic acid RNA; the use of retroviral gene insertion systems for human gene therapy is under investigation.

Reverse transcription: Synthesis of a single strand of DNA using an RNA template.

Ribo(se)nucleic acid (RNA): A molecule which resembles DNA in structure; acts as an adjunct in the execution and mediation of the genetic instructions encoded in DNA, for example, messenger RNA (mRNA) is transcribed from a single strand of a DNA molecule and is the template on which amino acids align to form the coded protein. RNA has been observed to function as an autocatalyst in transcription and possibly controls translation; RNA may have been the primordial genetic molecule functioning both as an enzyme and as a self-replicating repository of genetic information; RNA is the repository of hereditary information for some viruses (retroviruses).

Scale-up: The transition of a process from a laboratory-scale to an industrial-scale.

Sex cells: Cells which fuse together to form a fertilized egg—all other animal cells are called somatic cells, or body cells. In human beings the male sex cell is the sperm, and the female sex cell is the egg cell, also known as the ovum (pl. ova).

Sex chromosomes: Chromosomes that differ in number or morphology in different sexes and contain genes determining sex type; in human beings the sex chromosomal constitution of a normal female is XX and that of a normal male is XY.

Sex-linked (X-linked) trait: Determined by a gene located on a sex chromosome, usually the X chromosome in humans, and hence the term 'X-linked trait' is sometimes preferred.

Shotgunning (in microgenetic engineering): A technique for breaking up the entire genome of an organism into small pieces and then inserting those pieces into host cells, where they are cloned into a gene library for the organism.

Somatic cells: (Gr. *somes*, body.) Ordinary body cells as distinct from the reproductive cells or sex cells.

Speciation: Species formation; the creation of a barrier to interbreeding between individuals.

Species: (L. *species*, particular kind.) A group of individuals not able to breed with another such group; beyond being a taxonomic subdivision of a genus, the interpretation of what constitutes a species or speciation remains controversial; a group of closely related, morphologically similar individuals that actually or potentially interbreed but which maintain a genetic constitution different from all other such groups; a species is in reproductive isolation from other species because the hybrid offspring of such a union are infertile or inviable.

Subviral infectious elements: Nomadic nucleic acid molecules that are simpler than viruses in structure; they enter cells and multiply therein; viroids are an example.

Taxonomy: (Gr. *taxis*, arrangement; *nomos*, law.) The method of arrangement or classifying, particularly of living organisms. Taxonomic studies have led to the development of a system of classification which divides all living things into two large groups called kingdoms, each of which is divided into a series of major subgroups called phyla (sing. phylum). Each phylum is further divided into a series of successively smaller groups known as classes, orders, families, genera (sing. genus) and finally species. There is generally only one kind of organism in a species. In modern taxonomy, an organism is named using the binomial system, under which an organism's name is designated by the genus to which it belongs (generic name), followed by the name of the species (specific name). The name is always written with a capital letter and the specific name with a small letter; for example, the taxonomic name of humans is *Homo sapiens*. The Swedish naturalist Carl Linnaeus introduced the binomial system of naming organisms in 1735.

Teratogeny: (Gr. *teras*, monster; *genos*, birth.) The formation of monstrous foetuses or births; thalidomide was classified as a teratogenic drug.

Test-tube babies: Popular term for the resultant offspring of successful IVF, implantation, pregnancy and birth.

Toxin: A substance, in some cases produced by disease-causing micro-organisms, that is poisonous to other living organisms.

Transcription: The synthesis of an RNA molecule using a section of one strand of a DNA molecule as a template.

Transduction: The transfer of genetic material between cells mediated by an infectious mobile genetic element, for example a virus.

Transformation: The process whereby a piece of foreign 'naked' DNA is taken up from the surrounding medium by a cell in which it integrates and gives that cell new properties.

Translation: The conversion of the base sequence of a messenger RNA molecule (mRNA) into a polypeptide chain of amino acids for the construction of a particular protein.

Translocation (chromosomal): The displacment of part or all of one chromosome to another.

Transposable genetic element: A genetic element that moves from site to site within a cellular genome; now believed to be a major feature of all DNA (see MOBILE GENETIC ELEMENTS).

Vaccine: A substance introduced into the body in order to stimulate the activation of the body's immune system as a precautionary measure against future exposures to a particular pathogenic agent.

Vectors: Vectors are self-replicating entities used as vehicles to transfer foreign genes into living cells and then replicate and possibly also express them; examples are plasmids and viruses.

Vermes chaos: L. *vermis*, a worm; vermin, a noxious animal; animals destructive to game; animals injurious to crops and other possessions; noxious persons, in contempt; Gr. *chaos*, that confusion in which matter was supposed to have existed before it was reduced to order; confusion; disorder (see MICRO-ORGANISM).

Viroid: Piece of 'naked' (without a protein 'coat') infectious RNA.

Virus: A minute infectious agent; a mobile genetic element, composed of nucleic acid (DNA or RNA) wrapped in a protein coat, which can survive on its own but which cannot replicate; to replicate it must be inside a living cell; viruses are used as vectors.

Guide to Educational Resources in Genetic Engineering and Biotechnology

This guide is intended for students embarking upon a course of study which includes some content of genetic engineering or biotechnology. It includes references to books, articles and journals, practical experiments, computer software and audio-visual aids.

For students interested in evolution, *Conceptual Issues in Evolutionary Biology* (Cambridge, Mass. and London: MIT Press, 1984), edited by Elliot Sober, is an excellent selection of papers. *Darwin for Beginners* by Jonathan Miller and Boris Van Loon (London: Writers and Readers Publishing Cooperative Society, 1982) is an amusing illustrated introduction to Darwin's ideas on evolution. A good starting point for further study of the historical development of genetics in the twentieth century is Gunther Stent's *Molecular Genetics: An Introductory Narrative* (San Francisco: Freeman, 1971). The section of Stent's first chapter (pp. 1–29), which is on the birth of molecular biology, relates the transition from classical genetics (in which the gene was considered as an abstract concept) to molecular genetics, through the period of the quest for the physical basis of heredity. By the 1940s many physical scientists were directing their attention to the study of the nature of the gene, having been inspired by the suggestion of physicists such as Niels Bohr, that its elucidation might reveal hitherto undiscovered laws of physics. These ideas were popularized by Erwin Schroedinger in *What Is Life?* (Cambridge: Cambridge University Press, 1945), in which he takes a quantum approach to the phenomenon of mutation and discusses the nature of the gene, introducing the concept of the genetic code in its modern form. Although written just before the realization that DNA, rather than protein, carries the message of heredity, it remains an excellent book, and one which was an inspiration to James Watson and Francis Crick.

For a scientific history of the development of molecular biology and the personalities involved, Robert Olby's book, *The Path to the Double Helix* (London: Macmillan, 1974), provides a scholarly exposition. John Gribbin's *In Search of the Double Helix* (Reading: Corgi, 1985) is another historical account of the scientific events leading up to the cracking of the genetic code which stresses the importance of theoretical physics to the development of molecular biology.

Subtitled 'The Life and Work of Barbara McClintock', Evelyn Fox Keller's *A Feeling for the Organism* (New York: Freeman, 1983) poignantly describes the tenacious resistance of the community of orthodox geneticists to accept Barbara McClintock's evidence for the existence of mobile genetic elements. 'The Significance of Responses of the Genome to Challenge', *Science*, vol. 226, 16 November 1984, pp. 792–801, is a publication of the lecture Barbara McClintock delivered in Stockholm when she received her Nobel Prize in 1983. In this paper she reviews the experimental evidence for her interpretation of the genome.

A clear résumé of the science of genetics, incorporating a description of the techniques of microgenetic analysis as they are applied to the study of the genome, and including insights into the eucaryotic genome revealed by microgenetic analysis, is written by Maxine Singer of the National Cancer Institute in the USA, in 'The Genetic Program of Complex Organisms', *Frontiers in Science and Technology* (New York: Freeman, 1983), pp. 15–44. She concisely describes the role of mobile genetic elements in evolution, disease, gene control and development and differentiation. Andrew Scott's *Pirates of the Cell* (Oxford: Blackwell, 1987) is an excellent introduction to viruses and viral vaccines and Martin Day's *Plasmids* (London: Edward Arnold, 1982) is a concise and broad-ranging account of plasmids.

An easy-to-follow layperson's introduction to the science of genetics is *The Cartoon Guide to Genetics* (London: Harper & Row, 1984) by Larry Gonick and Mark Wheelis. An authoritative introduction to the techniques of genetic engineering can be found in *Recombinant DNA: A Short Course* (New York: Freeman, 1983) by James Watson *et al.* The comprehensive and well-illustrated description begins with basic background information in genetics and progresses to techniques of genetic engineering and their applications in research. A collection of articles which appeared in the journal *Scientific American* has been compiled and edited by David Freifelder in *Recombinant DNA: Readings from Scientific American* (San Francisco: Freeman, 1978), which includes an excellent article by Clifford Grobstein on the recombinant DNA debate. *Man-Made Life* (Oxford: Blackwell, 1982) by the zoologist Jeremy Cherfas is an introduction to the methods used in recombinant DNA technology. William Bains' book *Genetic Engineering for Almost Everybody* (Harmondsworth: Penguin, 1987) explains the techniques of genetic engineering in simple terms.

Books on the recombinant DNA debate are Clifford Grobstein *A Double Image of the Double Helix* (San Francisco: Freeman, 1979); *Genetic Alchemy: The Social History of the Recombinant DNA Controversy* (Mass: MIT, 1982) by Sheldon Krimsky; Jeremy Rifkin *Algeny: A New World* (London: Pelican, 1984); *The DNA Story: A Documentary History of Gene Cloning* (New York: Freeman, 1981) by James Watson and John Tooze, which includes reproductions of relevant newspaper and journal articles, documents and cartoons published at the time of the controversy. The US historian of science Susan Wright's article 'Molecular Biology or Molecular Politics? The Production of Scientific Consensus on the Hazards of Recombinant DNA Technology', *Social Studies of Science*, vol. 16, pp. 593–620 provides interesting insights into the way in which the consensus amongst molecular biologists regarding the safety of recombinant DNA work came about.

Good introductory cover to biotechnology can be found

in J.E. Smith's *Biotechnology* (London: Edward Arnold, 1981); and Steve Prentis *Biotechnology: A New Industrial Revolution* (London: Orbis, 1984), which includes an exposition of genetic engineering. A guide to further reading in biotechnology is *Biotechnology: A Review and Annotated Bibliography* (London: Frances Pinter, 1980) by Harry Rothman *et al.* Recommended for its concise introductory treatment is the 'Biotechnology Information Pamphlet', one of a series of pamphlets published by the American Chemical Society's Office of Federal Regulatory Programs, Department of Government Relations and Science Policy, with the aim of addressing public policy issues held to be of concern to both ACS members and the general public. The pamphlet provides a well-written account with diagrams, which in twelve pages covers basic cell biology and genetics, recombinant DNA technology, the applications of biotechnology using genetically engineered agents in the major industrial sectors, and an account of the arguments pitted in the debate over environmental release. A single copy of the pamphlet is available free of charge, and photo-duplication of the pamphlet is encouraged, provided proper acknowledgement is given. Additional copies are available for a charge from the office of Federal Regulatory Programs, ACS Department of Government Relations and Science Policy, 1155 16th Street, NW Washington, DC, 20036, USA.

Biotechnology and genetic engineering trade literature, which provides well-produced, though obviously uncritical, information on products, companies and processes, can be readily obtained by writing directly to companies or by using the business reply cards inserted in journals such as *Nature*. *Steam*, the occasional ICI Science Teachers Magazine, issue number 2, July 1984, pp. 19–21, has a feature entitled 'Biotechnology: A New Challenge to Schools', which includes a list of the names and addresses of some of the companies and organizations, such as Unilever, Tate & Lyle and Gist Brocades, which have published booklets on biotechnology. A single copy of *Steam* may be obtained free of charge. Further copies can be purchased from ICI Educational Publications, PO Box 96, 1 Hornchurch Close, Coventry, West Midlands, CV1 2QZ, England.

Books on the biotechnology industry are *The Gene Factory: Inside the Biotechnology Business* (London: Century, 1985) by the editor of *Biotechnology Bulletin*, John Elkington; and *Setting Genes to Work: The Industrial Era of Biotechnology* (London: Viking, 1985) by Stephanie Yanchinski, which concentrates on the pharmaceutical industry. *Biotechnology: International Trends and Perspectives* (Paris: OECD, 1982) by Alan Bull *et al.* contains a number of useful appendices relating to biotechnology and recombinant DNA technology.

The Biotechnology Information Service (BIS) (formerly the European Biotechnology Information Project (EBIP)) is based at the Science Reference Information Service (formerly the Science Reference Library) of the British Library, 9 Kean Street, London, WC2B 4AT. In addition to a free newsletter, it has publications on approximately fifteen specific areas, including culture collection and nucleic acid data banks, patents and forthcoming conferences, and responds to individual requests for information. Another recommended source of wide coverage is *Information Sources in Biotechnology* (Basingstoke: Macmillan) by Anita Crafts-Lighty (1987). Anyone interested in research into the social and economic implications of the diffusion of the new genetic engineering technology and biotechnology is eligible for membership of the Biotechnology Business Research Group (BBRG) and should contact Peter Wheale, School of Business, Oxford Polytechnic, OX9 1HX.

For regular up-to-date information, a weekly perusal of journals such as *Science*, *Nature* and *New Scientist* enables monitoring of major breakthroughs. There are approximately seventy non-technical journals with a significant biotechnology content. The BIS/EBIP has analysed these and presents the results—a breakdown of the content, cost and regularity—in a free publication entitled 'Keeping Up-to-Date in Biotechnology'. Included in this listing are seven or eight free journals, for example *Biobulletin*, the newsletter of SERCs Biotechnology Directorate, and *Genetic Engineering and Biotechnology Monitor*, a United

Nations Industrial Development Organization (UNIDO) publication, and the European Federation of Biotechnology Newsletter. Publishers' addresses are given in the BIS/EBIP list referred to above. For a modest subscription, *Industrial Biotechnology Wales* can be obtained from the Biotechnology Centre Wales, University College Swansea, Singleton Park, Swansea SA2 8PP, Wales. For rapid coverage of a large number of sources several journal abstracts and databases specializing in biotechnology are available, as well as several online host services. For further information contact the BIS.

We recommend the following publications on specific areas. *Genetic Prophecy: Beyond the Double Helix* (London: Granada, 1983) by the director of the US OTAs study of genetic engineering, Zsolt Harsanyi, and Richard Hutton, on the impact of the ability to screen for genetic disease and for individual susceptibility to environmentally induced disease. D.J. Weatherall's *The New Genetics and Clinical Practice* 2nd ed, (Oxford: Oxford University Press, 1985) is an excellent analysis of the likely practical benefits of the new techniques of molecular biology to the management of genetic disorders. The new reproductive technologies are described and analysed from an ethical and legal perspective in Peter Singer and Deane Wells's *The Reproduction Revolution*: *New Ways of Making Babies* (Oxford: Oxford University Press, 1984); and the publication of the Council for Science and Society, *Human Procreation*: *Ethical Aspects of the New Techniques* (Oxford: Oxford University Press, 1984). Edward Yoxen's *Unnatural Selection*: *Coming to Terms with the New Genetics* (London: Heinemann, 1986) discusses the ideology underlying the new genetic and reproductive technologies. The debate surrounding genetic determinism is argued from opposite sides in Richard Dawkin's *The Selfish Gene* (London: Paladin, 1982) and Steven Rose *et al.*, *Not in Our Genes* (Harmondsworth: Penguin, 1984). Sean Murphy *et al.* *No Fire, No Thunder* (London: Pluto Press, 1984) gives coverage with regard to chemical and biological warfare.

The *Recombinant DNA Technical Bulletin* (Office of Recombinant DNA activities, National Institute of Allergy and Infectious Diseases, NIH, 9000 Rockville Pike, Building

31, Room 3B 10, Bethesda, Md. 20205, USA) regularly publishes articles detailing the US regulatory powers and the activities of working parties with regard to specific areas. *Biotechnology: A Plain Man's Guide* (1986) is a concise compilation of information about the current UK regulations applicable to biotechnology and relevant contacts. It has been prepared by the Interdepartmental Committee on Biotechnology and published by the Department of Trade and Industry (DTI). For a free copy of this guide, contact the Biotechnology Unit, Laboratory of the Government Chemist, Cornwall House, Waterloo Road, London SE1 8XY. The Green Alliance is an organization which endeavours to increase environmental awareness through meetings, lobbying and publications. The Green Alliance publish a *Parliamentary Newsletter* every fortnight which contains abstracts from British newspapers and government proceedings on environmental issues. Their address is 60 Chandos Place, London, WC2N 4HG.

Edward Yoxen's *The Gene Business: Who Should Control Biotechnology?* (London: Pan, 1983) gives an accessible account and critical analysis of the commercialization of molecular biology in the 1970s and early 1980s together with some satirical cartoons. A critical review of biotechnology, which was funded by the European Foundation, is Edward Yoxen's 'Report of a Four-Nation Study of the Social and Environmental Impact of Biotechnology', *European Foundation*, 1987. Marc Lappe explores some of the ethical issues of the commercialization of recombinant DNA technology in *Broken Code* (San Francisco: Sierra, 1985). The journal *Science for People*, available from 25 Horsell Road, London N5 IXL, occasionally prints articles in this area. One of the few journals whose policy is to maintain a critical stance specifically towards genetic engineering is *GeneWATCH*, published by the Committee for Responsible Genetics (CRG), 186A South Street, Boston MA 02111, USA, from whom it is available at a modest subscription. Each issue includes a resources section, and the CRG has also produced two bibliographies of books, articles and films.

The report of the Royal Society, *Biotechnology and Education: The Report of a Working Group* (London: The

Royal Society, 1981) recommended that biotechnology should not be regarded as a separate subject, but that it should be integrated into existing curricula. A number of groups in the UK have drafted guides on the integration of biotechnology into existing science syllabuses. For further information, contact the Association for Science Education, Biotechnology Sub-committee, College Lane, Hatfield, Herts, AL10 9AA, England.

In the USA the Science through Science, Technology and Society (S-STS) Project is a national programme for pre-college education in STS funded by the National Science Foundation (NSF) and based at Pennsylvania State University. For information regarding the activities and educational resources of the S-STS Project in biotechnology and genetic engineering, contact S-STS Project, Pennsylvania State University, 128 Willard Building, University Park, PA 16902, USA.

The UK Microbiology in Schools Advisory Committee (MISAC) has produced a list of audio-visual materials for use in microbiology, copies of which can be obtained from the Executive Secretary, Institute of Biology, 20 Queensberry Place, London, SW7 2DZ, with an S.A.E. A list of local MISAC advisors for the teaching of microbiology in schools is also available from the address above. The Education Officer at the Institute of Biology (address above) has information regarding courses and careers in biological subjects.

There are a number of practicals which can be carried out by individuals with little scientific expertise or background for purposes of 'biotechnology appreciation', such as plant and animal tissue culture; monoclonal antibody blood typing; enzyme immobilization; the use of enzymes and micro-organisms in the home and in industry, for example in detergents, manufacture of fruit juices, cheese and yoghurt manufacture; and genetic studies, such as gene complementation, mutant characterization, transduction and restriction enzyme digestion analysis. Experimental kits and teaching aids, such as overhead projector transparencies, slides and film strips, and interactive computer programmes are available from Griffin and George, Gerrard Biological Centre, Worthing Road, East Preston, West

Sussex BN16 1AS, England, whose products include a genetic engineering kit; and Philip Harris Biological Ltd, Oldmixon, Weston-Super-Mare, Avon, BS24 9BJ, England, who together with Novo Industri have produced a set of overhead projection transparencies with teaching notes on enzymes in industrial and commercial use. Irena Olejnikowna has edited 'Biotechnology Resources for Schools and Colleges', *ASE Occasional Papers Series*, 1986, which includes 'recipes' for school use. *Molecular Cloning: A Laboratory Manual (New York: Cold Spring Harbor Laboratory, 1982) by Tony Maniatis et al.* is a cook book-style laboratory bench guide for more advanced experiments.

A number of educational computer programs have recently come onto the market which enable the simulation of biological experiments or help to explain scientific or technological principles. For example, computer software enables simulation of some of the decision-making required for the operation of a large-scale fermenter. *Fermt* is an example of an interactive computer program on the fermentation process based on the standard equation for microbial growth, the Monod equation; the student alters seven variables in the growth conditions with the aim of maximizing profit on the end-product. For further details contact the School of Biotechnology, Polytechnic of Central London, 115 New Cavendish Street, London, W1M 8JS. The *Biology/Science Materials Catalog*, published by the Carolina Biological Supply Company, 2700 York Road, Burlington, North Carolina 27215, USA, which includes listings of video programmes and computer software, is available free to full-time teachers and health instructors. Membership of the US National Science Teachers Association (NSTA), 1742 Connecticut Avenue, NW, Washington, DC, 20009, USA, includes subscription to one or more of NSTAs journals of science teaching—*Science and Children* (elementary/middle school), *The Science Teacher* (junior/ senior High School) and the *Journal of College Science Teaching* (college), which publish advertisements and reviews of teaching resources, and notices of NSTAs conferences and courses. Each January issue of all three magazines includes a supplement detailing science education suppliers, listed under equipment, media producers, educational services, computers/software and publishers.

As well as the *science* of genetic engineering, the available videos in this area portray the *business* of biotechnology. Recommended is the video by BBC Enterprises in *The Risk Business* series entitled 'A Licence to Breed Money' (1981). Also available on video cassette from the BBC is the BBC2 television series *Life Power* (1984), six programmes on biotechnology produced by Paul Kriwaczek. For details contact BBC Enterprises, Education and Training Sales, Woodlands, 80 Wood Lane, London W12 0TT, from whom a listing of BBC videos in this subject area can also be obtained; customers in the USA should apply to Films Inc., 773 Greenbay Road, Wilmette, Illinois 60091, USA. For a reasonable price, though less well edited, *The Bio-Bombshell* (1984), designed for older school pupils and college and university students, can be obtained from the publishers of *New Scientist* at New Scientist Quest, IPC Video and Cable, Room 2627, Kings Reach Tower, Stamford Street, London, SE1 9LS. The Open University Centre for Continuing Education short course PS621 on *Biotechnology* (1985) is designed to be accessible to the non-specialist. The six 25-minute television programmes linked with the course have been produced on video-cassette together with a similar quantity of supplementary material, and a special arrangement has been made to make the cassettes available to schools and colleges. Enquiries regarding the course and the programmes should be addressed to Margaret Swithenby, Course Co-ordinator for Biotechnology, Department of Biology, The Open University, Walton Hall, Milton Keynes, MK7 6AA. A video which conveys some of the social and economic impacts of one aspect of the commercialization of biotechnology, the use of sugar as a substitute for oil in the production of plastics, detergents and liquid fuel for cars, is 'Sweet Solutions' (BBC *Horizon*, 1979).

Fiction in the form of the novel or drama is an important collective experience and can be used effectively to communicate emotions and ideas and to explore issues and anxieties of universal concern. *The White Plague* is a novel by Frank Herbert published in 1982 by G.P. Putnam's Sons of New York, and considers the nefarious uses of genetic engineering in the hands of a mad scientist. The scientist in question purposely develops and releases an incurable pathogen but is shocked to realize that his engineered pathogen has undergone spontaneous genetic mutation in the natural environment. Frank Herbert, whilst drawing attention to the catastrophic potential of genetic engineering technology, portrays the utter futility of any conceivable attempt to limit or prevent its use. In *Rat and Super Rat* (1984), an unpublished play by the authors, Peter Wheale and Ruth McNally, the ethical issues and dangers encountered in the pursuit of microgenetic engineering for military purposes are explored through the medium of the theatre. Set in a laboratory in a military-industrial state, it depicts the discovery and development of the techniques to microgenetically engineer thermonuclear radiation-resistant human clones which can withstand thermonuclear radiation fallout with apparent impunity, and portrays the dilemma of the molecular biologist, Barbara Wass, upon realization of the consequences of her research work. For further details write to Ruth McNally, 27E Saint Andrew's Road, Bedford, MK40 2LL. Twentieth Century Fox *Warning Sign*, a film directed by Hal Barwood, depicts a biotechnology experiment gone awry which results in the mass-infection of laboratory workers with a dangerous virus. The film portrays military personnel conducting secret biological weapons research in a small town in Utah in the USA using an agrobiotechnology company as a cover. *Warning Sign*'s plot involves several breaches of safety protocol and questions the integrity of using biotechnology as a means for developing the agents of biological warfare.

List of Abbreviations

ABRC	Advisory Board for the Research Councils (UK)
ACGM	Advisory Committee on Genetic Manipulation (UK)
ADA	Adenosine deaminase deficiency
AGS	Advanced Genetic Sciences
AI	Artificial insemination
AID	Artificial insemination by a donor
AIDS	Acquired immune deficiency syndrome
AMRDC	Army Medical Research and Development Command (US)
APHIS	Animal and Plant Health Inspection Service of the USDA
ARRC	Agricultural Recombinant DNA Research Committee of the USDA
ATCC	American Type Culture Collection
BSCC	Biotechnology Science Co-ordinating Committee (US)
BSSRS	British Society for Social Responsibility in Science
BST	Bovine somatotropin
BTG	British Technology Group
CCPA	Court of Customs and Patent Appeals (US)
CDC	Centers for Disease Control (US)
CERB	Cambridge Experimentation Review Board
COGENE	Committee on Genetic Experimentation
CMUBR	Committee on the Military Use of Biological Research (US)
CND	Campaign for Nuclear Disarmament
CRG	Committee for Responsible Genetics (US)
CSM	Committee on Safety of Medicines (UK)
CVS	Chorionic villus sampling
DES	Department of Education and Science (UK)
DES	Diethylstilboestrol
DHHS	Department of Health and Human Services (US)
DHSS	Department of Health and Social Security (UK)
DNA	Deoxyribo(se)nucleic acid

cDNA	copy DNA
DoD	Department of Defense (US)
DoE	Department of the Environment (UK)
DPAG	Dangerous Pathogens Advisory Group (UK)
DPC	Domestic Policy Council (US)
DTI	Department of Trade and Industry (UK)
EC	European Commission
EEC	European Economic Community
EIS	Environmental Impact Statement (US)
EMBL	European Molecular Biology Laboratory
EMBO	European Molecular Biology Organisation
EPA	Environmental Protection Agency (US)
EVIST	Ethical Values in Science and Technology (US)
FCCSET	Federal Coordinating Council for Science, Engineering and Technology (US)
FDA	Food and Drugs Administration (US)
FIFRA	Federal Insecticide, Fungicide and Rodenticide Act (US)
GAO	General Accounting Office (US)
GMAG	Genetic Manipulation Advisory Group (UK)
HIV	Human immunodeficiency virus
HLA	Human leucocyte antigen
HPRT	Hypoxanthine-guanine phosphoribosyl transferase
HSC	Health and Safety Commission (UK)
HSE	Health and Safety Executive (UK)
HSW Act	Health and Safety at Work Act 1974 (UK)
HTLV	Originally, human T-cell leucaemia virus; changed to human T-cell lymphotropic virus
IBA	Industrial Biotechnology Association (US)
IBC	Originally, Institutional Biohazard Committee; changed to Institutional Biosafety Committee (US)
INA-	Ice-nucleation inactive
IVF	*In vitro* fertilization
LAV	Lymphadenopathy associated virus
MAFF	Ministry of Agriculture, Fisheries and Food (UK)
MIT	Massachussetts Institute of Technology
MoD	Ministry of Defence (UK)
MRC	Medical Research Council (UK)
NAS	National Academy of Sciences (US)
NATO	North American Treaty Organization
NCC	Nature Conservancy Council (UK)
NEB	National Enterprise Board (UK)
NEPA	National Environmental Policy Act (US)
NIAID	National Institute for Allergic and Infectious Diseases (US)
NIBSC	National Institute for Biological Standards and Control (UK)
NIH	National Institutes of Health (US)

NIH-RAC	Recombinant DNA Advisory Committee of the NIH (US)
NIOSH	National Institute of Occupational Safety and Health (US)
NRDC	National Research Development Corporation (UK)
NRDC	Natural Resources Defense Council (US)
NSF	National Science Foundation (US)
ORDA	Office of Recombinant DNA Activities (US)
OSHA	Occupational Safety and Health Administration (US)
OSTP	Office of Science and Technology Policy (US)
OTA	Office of Technology Assessment (US)
PAHO	Pan American Health Organization
PKU	Phenylketonuria
PNAS	Proceedings of the National Academy of Sciences (US)
PNP	Purine nucleotide phosphorylase deficiency
PTO	Patent and Trademark Office (US)
RAC	Recombinant DNA Advisory Committee of the NIH (US)
RFLP	Restriction fragment length polymorphism
RNA	Ribo(se)nucleic acid
mRNA	messenger RNA
SIPRI	Stockholm International Peace Research Institute
SMO	Supervisory Medical Officer (UK)
SPUC	Society for the Protection of Unborn Children
SV40	Simian virus 40
TMV	Tobacco mosaic virus
TSCA	Toxic Substances Control Act (US)
UCLA	University of California, Los Angeles
UCSF	University of California, San Francisco
UN	United Nations
UPOV	International Union for the Protection of Plant Varieties 1968
USDA	United States Department of Agriculture
USDA-APHIS	Animal and Plant Health Inspection Service of the USDA
USDA-ARRC	Agricultural Recombinant DNA Research Committee of the USDA
VPC	Veterinary Products Committee (UK)
WHO	World Health Organisation
YAC	Yeast artificial chromosome

Name Index

320

Subject Index